Bovine Somatotropin and Emerging Issues

Bovine Somatotropin and Emerging Issues

An Assessment

EDITED BY
Milton C. Hallberg

LONDON AND NEW YORK

First published 1992 by Westview Press, Inc.

Published 2018 by Routledge
52 Vanderbilt Avenue, New York, NY 10017
2 Park Square, Milton Park, Abingdon, Oxon OX14 4RN

Routledge is an imprint of the Taylor & Francis Group, an informa business

Library of Congress Cataloging-in-Publication Data
Bovine somatotropin and emerging issues : an assessment / edited by
Milton C. Hallberg.
p. cm. — (Westview special studies in agriculture science and policy)
Includes bibliographical references and index.
ISBN 0-8133-0603-5
1. Bovine somatotropin. 2. Dairy farming. 3. Dairy products.
4. Animal biotechnology. I. Hallberg, M. C. (Milton C.)
II. Series.
SF768.2.C3B68 1992
363.19'29—dc20 91-19136
CIP

ISBN 13: 978-0-367-00557-3 (hbk)
ISBN 13: 978-0-367-15544-5 (pbk)

Contents

Foreword

This book prepared by Professor Hallberg and his collaborators is, by the standards of twentieth-century agricultural science, as unusual an undertaking as it is a welcome one. This volume is welcome for obvious reasons: It represents a concerted effort to report objective information, primarily for an audience of nonspecialists, on a scientific-technical controversy--the BST debate--that has been among the most politicized and protracted in post-World War II agricultural science. The editor has called on the expertise of a number of social and natural scientists who do not have an immediate professional, financial, or ideological stake in these debates. Each of the chapters, most of which are essentially review articles, has an important message. The authors are by no means of a single mind; in as wide-ranging a debate as that over BST, there is unlikely to be a single formula for objectivity. But the authors share a conviction of detachment that has not ordinarily been the case in the literature on this topic.

This book is unusual in the fact that it had to be undertaken in the first place. Scientific and technical controversy is, of course, normal, since the work of most research scientists is focused on learning things we do not yet know or are not sure about. Disagreements at the frontiers of our knowledge are to be expected. But most scientific controversies are typically confined to the scientific community and insulated from public view. The BST case, however, has been one in which the debate has extended far beyond the scientific community, coming to be a high-visibility public policy debate in virtually all advanced industrial countries. Moreover, the BST debate has not simply been a debate within science. It has come to be a debate about science as well. That is, disagreements over BST have been heightened to a point where some are no longer confident that scientists and scientific institutions can play a meaningful role in resolving them. There has, in other words, been a trend toward erosion of trust in the ability of public scientists to be neutral and dispassionate arbiters of conflicting views on science and technology issues.

The breadth of the BST controversy suggests that one must begin by understanding its roots, within the emergence of what can be called the "new politics of agricultural research." During the heyday of American public agricultural research, from the end of World War II to the end of the Vietnam War, the public research system was, most fundamentally, a state-government-supported system mainly oriented to the technical needs of traditional state-based constituent groups, the organized commodity groups in particular. The aggregate impact of the land-grant system was one of sustained productivity increase across the length and breadth of the country. The system was praised for its productivity accomplishments, and ensuring long-term productivity increase was mainly what was expected of it.

This was all to change, beginning in the 1970s, in two quite different ways. First, there emerged a set of forces--propelled initially by scientists and scientific organizations wishing to move agricultural research in a more *basic* direction and later by private agribusiness firms and federal science policymakers--that has had the effect of encouraging agricultural scientists and administrators to emphasize basic research and national competitiveness goals over traditional in-state client service. Second, and nearly simultaneously, public interest groups began to criticize the public agricultural research system for what they saw as a slavish pursuit of productivity increases and for having ignored the deleterious effects of agricultural technology on the rural environment, family farmers, and rural communities.

The result is that land-grant institutions now find themselves in a *triangular squeeze*: state-based commodity groups want to preserve their traditional prerogatives; they want *their* state agricultural experiment station to continue to devote its efforts to providing locally adapted technology geared to local agroecological conditions and the technical needs of producers in the state. Agribusiness firms, basic-science-oriented groups, and federal science policymakers want the land-grant system to pursue more basic or generic research, most of which is ultimately to be transferred to the private sector to bolster the American competitive position in "high-technology." Public-interest groups (along with growing numbers of farm and rural people who share the goals of these groups) want the land-grant system to be more responsive to the concerns of family farmers, small farmers, organic farmers, consumers, and so on, and to redress the impacts of agricultural technology on rural communities and the rural environment.

Several aspects of the new triangular squeeze should be stressed. First, each of the land-grant constituents at each corner of the triangle is a legitimate one. Second, it is, I believe, a reality of modern public agricultural research that while the three corners of the triangular squeeze are not equal in power, each set of constituents has the capacity to do considerable damage to the land-grant system if its concerns do not receive consideration. Third, while there is room for research policy decisions to satisfy two or more constituent categories simultaneously (e.g., devoting biotechnology research to developing locally adapted technology), there are significant respects in which these groups' interests are contradictory. Fourth, the main imperative of the triangular squeeze is that land-grant officials will have little recourse but to strive toward institutional neutrality and to see their research, extension, and teaching programs as a portfolio in which the concerns of each constituent group are amply represented. Finally, the triangular squeeze has crystallized during a time in which most land-grant universities, colleges of agriculture, and experiment stations are facing stagnation of federal funding and, in most instances, stagnant if not declining state funding. Land-grant fiscal stress has made acquisition of extramural funding, both corporate and public, more attractive. There is, then, an understandable tendency to orient research to the goals of those who pay for it at the same time that fiscal stress has made acquisition of extramural funding, both corporate and public, more attractive. This has limited the ability of administrators to accommodate new constituents through expansion of resources and addition of new programs.

The 1980s was a period of massive change in colleges of agriculture and agricultural experiment stations. These changes can be understood mainly with reference to the triangular configuration of the land-grant system's constituents. The land-grant system slowly but surely reduced its resources devoted to applied, locally adapted research, while attempting to increase the efficiency and visibility of technology transfer, through extension, in order to compensate. The land-grant system has very substantially increased its commitment to biotechnology, to satisfy simultaneously its basic science critics, federal science policymakers, and its agribusiness constituents. The colleges of agriculture and experiment stations have made some gestures toward accommodating the concerns of public-interest groups and farmer critics.[1]

BST is a biotechnology, in the sense that recombinant BST is produced through industrial microbiology from a genetically engineered organism and research on its efficacy involved some sophisticated animal biochemistry. It is also essentially a generic technology, which is to say that it is applicable almost anywhere. BST thus typifies much of the current generation of agricultural-biotechnology research in which generic applications are given emphasis over locally adapted ones. I suspect that one of the reasons the BST controversy became so heated is that BST for many farmers symbolized the shaky commitment of the land-grant system to an emphasis on state-oriented research. While dairy farmers were by no means in the vanguard of opposition to BST, many farmers became concerned and actively or passively supported efforts to curb the technology. Further, agricultural biotechnology has been disproportionately important, relative to its role in the world biotechnology industry, in stimulating and shaping public debate. Probably only about 30 percent of U.S. biotechnology firms do significant research on agricultural or food applications, and less than 20 percent of the total U.S. biotechnology investment is geared to agriculture. Yet many of the most controversial aspects of biotechnology (environmental implications of the release of genetically engineered organisms for the benefit of agricultural crops; the most recent extensions of intellectual property restrictions on life forms; food safety and regulatory concerns; industrial concentration issues) pertain primarily, if not exclusively, to agriculture and food. Agriculture and food are sensitive matters. Everyone must eat; it is therefore not

---

[1]The public-interest agenda has over the past decade increasingly taken on a "green" cast as issues of agricultural sustainability, the agricultural environment, food safety, technical risk, and the like have come to predominate over the social-justice issues emphasized in the preceding decade. This agenda has been moved forward by federal legislation, particularly that in recent farm bills (e.g., their LISA and Conservation Reserve Program provisions). One of the most important implications of the "greening" of opposition to the land-grant system is that these concerns are increasingly expressed in scientific terms (e.g., criticism of agricultural chemicals on agroecological grounds, or of BST animal-health data on methodological grounds). This trend accounts for the fact that public agricultural research policy debates increasingly take the form of debates within science--and often become full-blown scientific and technical controversies.

surprising that so many consider the conditions under which our food is produced, processed, and distributed to be their business.

How, then, are we to understand and go forward from the BST controversy? I believe there are several lessons to be learned from how this scientific controversy has unfolded.

One set of lessons is essentially political and can be seen most starkly if we keep in mind that when the controversy erupted the strongest boosters and critics of BST were in full agreement that this technology would be an immediate, startling, "revolutionary" impetus to productivity increase in dairying. BST's corporate boosters saw this technology as the first of a new generation of agricultural inputs to be derived from applying the sciences of molecular and cell biology. Shortly thereafter BST became one of the darlings of the biotechnology industry, even among firms with no agricultural-biotechnology research; the biotechnology industry at this time was still in a frenzied scramble to impress investors, attract venture capital, build new life sciences research centers, and so on, and the BST breakthrough was often employed to demonstrate that biotechnology is already a concrete reality. BST's scientific advocates no doubt saw it, at least in part, in terms of professional advancement. BST's land-grant administration boosters seized upon the technology's promising productivity implications as a means to demonstrate to state and federal government funders how appropriating more money for new land-grant biotechnology programs would improve the economic competitiveness of their states and the nation as a whole; they generally did so without much thought being given to how news of these scientific miracles and technological "revolutions" would play among public-interest groups and among dairy farmers in the depths of a devastating, over-production-induced, mid-1980s farm crisis.

BST's critics ultimately repeated what they had heard from BST proponents: that the technology would lead to output increases of 25 percent or more, to feed efficiency improvements nearly as large, and to staggering losses in the number of dairy farmers. We now know, after more careful analysis and sober reflection, that these early data so confidently disseminated through news releases, testimony at legislative hearings, institutional research funding proposals, and biotechnology program brochures were exaggerated in many respects, as noted in several of the chapters in this collection. Though many scientists would become more guarded in their estimates of the ultimate productivity impacts of BST, both the technology boosters and detractors tended to persist in painting an unrealistic portrait of this technology.

In my view, there are several interrelated political lessons of the BST episode. Land-grant scientists and administrators must continue to learn that the role of public agricultural research is not simply to unleash technological revolutions. While productivity increase is a legitimate and necessary goal of public research, it is not its only goal. Technology-generating institutions ultimately must bear responsibility for the full range of impacts of new technology, including the negative or adverse ones, and recognize that those persons displaced are just as much constituents of the system as those who use new technology to personal advantage. The triangular squeeze of the land-grant system means that it must be more sensitive to a wider range of concerns, recognize the pluralism of its legitimate constituents, and strive for institutional neutrality.

A further lesson from the BST controversy is that we must yet deal with the reality that agricultural biotechnology remains both symbol and substance. Agricultural biotechnology, including but not limited to BST, is obviously a material reality that will affect farm and rural people for decades into the future. But biotechnology is also a symbol of, depending upon one's point of view, all that is good and all that is bad in the land-grant system. BST has become a lightning rod for agricultural-biotechnology issues in general and most importantly for issues that were submerged before the development of recombinant BST and that transcend BST (e.g., public involvement in research; recognition that the constituents of experiment stations include more than just commodity groups, groups in a position to fund research, and firms with a commercial interest in research results). I am increasingly of the opinion that the compulsiveness with which various groups are lining up in support of or opposition to biotechnology on essentially symbolic grounds is leading to an exaggerated "scientization" of debate about what should be the future of American agriculture. Not all positive and negative dimensions of American agriculture are matters of technology. Institutions and policy make a great deal of difference. But institutional and policy issues are often obscured when we focus so narrowly and fervently on science and technology. Those who aspire to an agriculture that is satisfactorily competitive internationally should recognize that whether or not BST is developed or approved for use here will be a trifling matter in the long term. Those who aspire to a more environmentally benign or a more just agricultural system must recognize as well that, in general, the configuration of new technologies derives more from social institutions and policy as a whole than from the immediacies of public agricultural research policy.

A third lesson from BST is a broader national research and regulatory policy matter. Governments play a dual--and, more often than not, a contradictory--role in generating new agricultural technologies: They serve simultaneously as promoters and regulators of new technology. This is, in a sense, a clear, but obviously unavoidable, institutional conflict of interest. Governments in general, including public research institutions at the front line of technology development, must take both responsibilities seriously.

We live in an era of "scientization of values"--an era in which ethical and value issues are being pursued through rival applications and interpretations of science--and of the increasingly transparent value-ladenness of science. Some may lament the passing of the pre-biotechnology era when one's existence as a land-grant scientist or administrator was simpler than it has now become. But this new era, in which the connections between ethics, values, and social goals on the one hand, and research and technology on the other are being laid bare, has great promise in helping land-grant institutions to make wise choices. In this regard, there is much to be learned from the BST issues of the past half dozen or so years.

I look upon this book as an effort to build or restore the stature of land-grant colleges of agriculture as, above all, institutions to which the public can look for objective information. To supply this information becomes a very demanding task when scientific controversies become full-blown technical and social policy struggles. The editor and contributors show that this goal is still within reach. It may even prove to be that the success of the effort will be gauged by the range of persons who find fault with it.

Frederick H. Buttel
Professor of Rural Sociology
Cornell University

Preface

BST is one of the first of the agricultural technologies to result from recombinant-DNA technology and is thus capable of being produced through industrial microbiology. Like all previous agricultural technologies, BST promises desirable, primary consequences as well as some unwanted, secondary impacts. Unlike some previous technologies, though, there is in the minds of some consumers uncertainty as to the safety of milk produced from cows that have been administered BST, and uncertainty in the minds of others as to the consequences for the health and welfare of the cow that is to be subjected to injections of the product.

BST is also the first major test-case biotechnology. How society reacts to BST will probably set a precedent for how society reacts to similar technologies already available (e.g., porcine somatotropin) or yet to be developed and tested. Will special imperatives be invoked and specific legislation be formulated to prevent anticipated unwanted consequences of the adoption and use of these technologies? Or will society treat these technologies as any other, relying on existing food safety and animal health regulations to deal with the unwanted consequences, and relying on the marketplace to guide adoption and use decisions?

Policy choices must not be made in a vacuum. The public must know about the desirable as well as undesirable impacts of these technologies in order to make informed judgements leading to rational policy choices. Accordingly, this book offers a full discussion of the issues surrounding BST and a comprehensive review of research findings on the probable consequences of BST, highlighting what is currently known and unknown about this technology. It is targeted to scientists and science managers in diverse fields, as well as to non-technical audiences. It will be of interest to the agricultural industries, to the press, to educators associated with science, technology and society (STS) programs, to professionals in the life sciences interested in becoming more broadly informed about BST and its impacts, and to faculty administrators who wish to become updated on BST and its full implications for the dairy industry and related sectors. The book will follow a generalized framework within which biotech issues will be examined. Here BST provides the case study, but the approach used offers a blueprint for similar studies of other biotechnological developments. In general, the book is aimed at *presenting the facts* rather than on articulating a definitive position on BST.

The first section of the book sets the stage for the remainder of the chapters by considering the general social and ethical issues involved in biotechnology. The authors of the first chapter examine the emerging developments in agricultural biotechnology, of which BST is one. They also review the policy issues these developments raise and outline broad potential impacts of these technologies on science, farmers and rural communities, consumers, and the international community. The second chapter of this initial section examines the ethical issues raised by BST. The author of Chapter 2 outlines the sources of ethical conflicts, examines the important contribution that risk and uncertainty have made to the BST debate, and discusses the prospects for political consensus in the face of the dominant ethical issues.

The second section of the book reviews what research has revealed about the technical and managerial aspects of BST: cow yield responses, animal nutrition requirements, animal health considerations, and dairy herd management. The third section reviews (a) what is known or expected concerning the future adoption of BST by the nation's dairy farmers, and (b) the probable economic consequences of full adoption of BST on the dairy and related agricultural sectors. The fourth section examines more-general market consequences of BST adoption: safety and quality of milk and dairy products produced from milk of BST-treated cows, probable consumer acceptance or rejection of dairy products when BST is used, expected long-term consequences of full adoption of BST on farm and retail prices of milk and dairy products and on regional competitiveness, and the salient international consequences of BST adoption.

A project such as this is clearly the product of a large number of dedicated professionals. It would not have happened, of course, had the authors and co-authors not been willing to put considerable time and effort into preparing the chapters and responding to editors' and reviewers' comments. It would not have been as high quality a product without the efforts of the several chapter reviewers. I wish to thank each and every one of these individuals for their contributions.

Glenn Bengtson served as technical editor for the project. He pleasantly and professionally edited the entire manuscript and provided invaluable guidance through the whole process. It was a great pleasure to hear nearly every contributor be complimentary toward his editorial work. Glenn also assisted with several other tasks related to preparing the final product. Clearly the book has benefitted immensely from his association with the effort. Glenn merits a special thanks.

Jane Mease expertly and professionally handled all the word-processing tasks required to convert the manuscripts from the authors' word processors to the final typeset form. She was not satisfied just to get it done. She was concerned as well about doing it right, and she was willing to go the extra mile to do it right. Jane also merits a special thanks. Tura Eisele attended to all the details associated with developing the graphic material included in the chapters of the book.

Finally, I am grateful to the Department of Agricultural Economics and Rural Sociology and to the Research Office of the College of Agriculture at The Pennsylvania State University for their encouragement and financial support of this project.

M. C. Hallberg
Professor of Agricultural Economics
The Pennsylvania State University

PART ONE

Biotechnology and Society

1

Emerging Trends, Consequences, and Policy Issues in Agricultural Biotechnology

William B. Lacy, Laura R. Lacy, and Lawrence Busch

The world economy is currently undergoing major restructuring and reorganization. A central factor in these changes is the development and diffusion of fundamentally new kinds of technologies, in particular, computers and the new biotechnologies. Social and economic changes and impacts stimulated by these enhanced capacities in science and technology are beginning to emerge in every area of human life, including health, transportation, communication, agriculture, and the food supply. Scientists, government officials, and industry leaders are particularly optimistic about the new biotechnologies. The Congressional Office of Technology Assessment (OTA) views the United States as "at the brink of a new scientific revolution that could change the lives and futures of its citizens as dramatically as did the industrial revolution two centuries ago and the computer revolution today. The ability to manipulate genetic material to achieve specific outcomes in living organisms (and in some cases their offspring) promises major changes in many aspects of modern life" (Office of Technology Assessment 1987:9).

University, government, and corporate leaders are equally optimistic that the new agricultural biotechnologies will continue the evolution of agricultural technology in the United States. Recently, the president of a large U.S. public university noted, "Biotechnology, genetic manipulation and engineering research will have tremendous impact on the crops and animals we grow for food, affecting agriculture in ways never before dreamed possible" (Lacy et al. 1988). Monsanto (1989), a large multinational corporation, predicted that biotechnology would revolutionize farming in the future with products based on nature's own methods, making farming more efficient, reliable, environmentally friendly, and profitable for farmers. Moreover, plants will be given the built-in ability to fend off insects and disease and to resist stress, animals will be born vaccinated, pigs will produce leaner meat and grow faster, cows will produce milk more economically, and food crops will be more nutritious and easier to process. The Monsanto statement concluded that biotechnology offers the farmer an opportunity to retain a competitive edge by producing crops and livestock of higher quality at lower cost.

Yet, as with most new technologies, apprehension and concern exist about the technical, environmental, economic, social, and value implications of the use of biotechnology in agriculture. Each change is associated not only with new benefits, but also with new risks, latent complications, and long-term consequences, often poorly understood.

This chapter examines the broad emerging trends, consequences, and policy issues surrounding the development of agricultural biotechnology in the public and private sectors. Within this context, the remaining chapters will examine, in depth, an important and early application of one of the products of these technologies (bovine somatotropin, or BST) to the U.S. dairy industry. A recent OTA report (Office of Technology Assessment 1991:3) states "The dairy industry will lead U.S. agriculture into the biotechnology era of the 1990s and will also feel the first profound impacts of emerging technologies." While many of the new products and processes may create some controversy, it is likely for the immediate future that the most pervasive and controversial application will be the use of BST produced through recombinant-DNA (rDNA) technology. Research has demonstrated that BST can boost milk yield per cow 10 to 15 percent and feed efficiency 5 to 10 percent under commercial conditions, but concerns have been raised about food safety for humans, stress on and safety of the cows and the economic consequences of BST's use for the industry.

Consumer advocates have raised questions about the quality and safety of milk produced using BST. In response to these concerns two states placed a one-year moratorium on the use of BST, that would continue even in the face of subsequent Food and Drug Administration (FDA) approval, and up to four states are seriously considering laws that would require milk and dairy products produced from BST-supplemented cows to be so labeled. Other issues surrounding this technology include the appropriateness of public-sector investment in its development, the relative lack of knowledge about the benefits and the risks of its application, and ultimately, questions about which and whose social needs are being met by this and other new technologies.

Since biotechnology, particularly in agriculture, is still in its infancy, the lack of applications and products allows us to examine alternative paths open to development. As Winner (1986:29) has observed, "by far the greatest latitude of choice exists the very first time a particular instrument, system or technique is introduced." It is likely, therefore, that focusing on this controversial and potentially influential technology at this stage of its development can be helpful to scientists, research administrators, policymakers, and the public.

Public Perceptions and Predictions

Biotechnology and genetic engineering have already begun to permeate our thinking, if not our everyday lives. They are also not without controversy, generating major debates and disagreements. Citizens in the United States are being increasingly exposed to information about biotechnology, particularly genetic engineering. A limited number of surveys have been conducted within the last five years to assess the public's awareness, understanding, and perception of the impact of genetic engineering. The most comprehensive is the 1987 national survey by the Office of Technology Assessment (1987).

These studies indicate that the public's interest in and knowledge of biotechnology, in general, and agricultural biotechnology, in particular, increased throughout the 1980s. While there is public concern about genetic manipulation, various applications involving no risk to humans or to the environment enjoy widespread acceptance. A majority of the citizens surveyed approve of the environmental use of genetically engineered organisms and field tests in their own communities. Moreover, roughly two-thirds of the public believe that genetic engineering will be more beneficial than detrimental (Lacy et al. 1991).

Public support for biotechnology, however, is tempered by major reservations. Two-thirds of America's citizens see a likelihood of environmental risk from genetically altered bacteria, and nearly half see a similar risk from genetically engineered plants and animals (Office of Technology Assessment 1987). Even more have expressed concern about food safety; more than three-quarters of the public expressed concern about eating genetically engineered food, particularly genetically engineered meat or dairy products. Comparably large percentages of the population appear worried about the negative socioeconomic consequences for farmers (Hoban 1990). Finally, while most believe application of these new technologies to plants and animals is appropriate, one in four believe that even genetic manipulation to create hybrid plants and animals is morally wrong (Office of Technology Assessment 1987). As a result of these concerns and beliefs, the majority of the public also believes that, while research should continue, strict regulation is necessary.

Given this general public concern, development of meaningful discussions and decisions regarding biotechnology, will require an awareness of public perceptions concerning who are credible spokespersons to raise and discuss the issues, and who are appropriate decision makers to determine the products, processes, and regulations of biotechnology. The public was asked who it believed most credible in providing information about the new biotechnologies. In the OTA survey (1987), 86 percent of the public indicated it is most likely to believe university scientists regarding risks of genetically engineered products. However, confidence in these scientists may be decreasing, as reflected in the general decline in public confidence in science and technology over the last three decades. The public also views public health officials and environmental groups as highly credible sources of information. The public would even believe these groups over federal agencies, particularly when these groups' opinions conflict with the federal agencies. The least amount of confidence resides in the media and the companies and corporations that are developing the new products and processes (Lacy et al. 1991).

Equally important is the question of who should decide upon the products and processes to be developed and the specific regulations to govern commercial applications. The public overwhelmingly believes that strict regulation of technology is necessary because it perceives potential dangers from these innovations. A government agency is the most frequently cited choice (37%) for defining commercial applications. An external scientific body was preferred by 29 percent. Only 13 percent felt that the decision should be left to the company that developed the product (Office of Technology Assessment 1987).

Biological Developments

Despite these broad public concerns, and the optimistic predictions of scientists, the government, and industry, most of the products, processes, and impacts of the new biotechnologies, (particularly in agriculture) remain promises for the future. Even in the pharmaceutical industry, where biotechnology has been an integral part of research for more than a decade, only a small number of genetically engineered drugs have been approved by the Food and Drug Administration for human prescription use (Holden 1990).

Diagnostic and therapeutic applications in the human health field continue to be the focus of most research and development investments in biotechnology. Agriculture, as a field of application for industrial biotechnology, is a distant third. Indeed, a recent report of the Organization for Economic Cooperation and Development (OECD), entitled *Economic and Wider Impacts of Biotechnology*, concludes that not until the turn of the century will biotechnology begin to play an economic and social role comparable to that of information technology, and only in the second decade of the century will it have major macroeconomic impacts (Dixon 1989).

Nevertheless, during the past 15 years a series of important discoveries and dramatic developments have emerged from basic research in molecular biology. After the first successfully directed insertion of foreign DNA into a host microorganism in 1973, researchers recognized the opportunities for advances in basic molecular biology, and the commercial potential in such areas as pharmaceuticals, animal and plant agriculture, specialty chemicals and food additives, commodity-specific chemicals, energy production, and bioelectronics. This initial success was quickly followed in 1974 by the expression in bacteria of a gene cloned from a different species. Other major biological developments in the last decade have included the first rDNA animal vaccine approved for use in Europe (1981), the approval of the first rDNA pharmaceutical product (human insulin) for use in the United States and the United Kingdom (1982), the first expression of a plant gene in a plant of a different species (1983), and the 1988 transfer of genes from other species into mice creating the first transgenic animal (Wheeler 1988). In mid-1989, researchers at the National Institutes of Health (NIH) introduced, for the first time, a foreign gene into a human patient in an experiment designed to develop a new way of treating malignant melanoma (Associated Press

1989). Recently, human hemoglobin, successfully produced in transgenic cattle and transgenic swine, was tested in human clinical trials (Hilts 1991).

Animals

In agriculture, the impacts of molecular biology and biotechnology may entail major changes in the traditional means of producing animals, plants, and food products. Since World War II, the animal sciences have benefited from a molecular biology emphasis in the health sciences; the results of medical research were often directly transferrable to meat and dairy animals. Future applications of biotechnology are likely to enhance livestock production through better diagnostic products, vaccines, growth promotants, and manipulation of the animal genome to achieve expression of desired traits. Among the recent developments are a genetically engineered vaccine against scours (a yearly killer of millions of newborn calves and piglets), a subunit vaccine engineered to prevent foot-and-mouth disease, the pseudorabies vaccine, and BST, the topic of this book, a genetically engineered hormone effective in increasing per-cow milk production (Martin and Baumgardt 1991).

Other animal science research focuses on reducing the fat content of farm animals; genetic engineering is one of several approaches employed. Transgenic animals with elevated growth hormone levels were shown to increase muscle growth more efficiently and to reduce fat production without compromising nutritional value (Sun 1988). Improvements in muscle rate gains (10 - 45%), feed efficiency (15 - 35%), and subcutaneous fat reductions (15 - 70%) have been reported. However, long-term elevation of bovine growth hormone (bGH) in two lines of transgenic pigs also resulted in several negative effects, including gastric ulcers, arthritis, cardiomegaly, dermatitis, and renal disease (Pursel et al. 1989).

A promising recent development in animal biotechnology is the production, in cattle and swine, of human hemoglobin, the oxygen carrying component of blood. In this work researchers injected copies of the two genes for human hemoglobin into day-old pig embryos (Hilts 1991). The goal of this research is a genetically engineered hemoglobin that can be mass produced as an inexpensive synthetic blood. Such a synthetic blood would be free of viruses, such as the

AIDS and hepatitis viruses, that may contaminate human blood. Synthetic blood could be given to a person without concern for blood type, and would have a longer shelf life than whole blood. However, serious questions arise about the safety of using hemoglobin produced by animals for human transfusions. UpJohn recently confirmed that it has suspended human trials with a bovine-produced hemoglobin blood substitute, due to some unexpected medical events in early testing (Hilts 1991). Among the questions are whether animal viruses might escape the purification process and cause illness in humans and whether cellular debris missed in purification might cause kidney damage or allergic reactions. Consequently, these early promising technical developments in animal biotechnology, including BST for dairy cattle, illustrate that even apparently successful and well-intended biotechnology applications can produce a range of both positive and negative biological and social effects.

Plants

Although the earliest advances were in the animal sciences, many analysts predict that the new biotechnologies may have their greatest impact on the plant sciences. Monsanto scientists note that the potential of genetically engineered plants, for exploring unique aspects of gene regulation and development and for producing new commercial crop varieties, has generated much excitement among scientists and has propelled research into many new areas (Gasser and Fraley 1989). The current work in plant biotechnology emphasizes the modification of specific plant characteristics (e.g., resistance to weeds, pests, herbicides, and pesticides; tolerance to stress; and improved nutritional content), and of traits of microorganisms that could be important to plant agriculture (e.g., those that foster pest resistance, nitrogen fixation, frost resistance, and disease suppression). Initially, research has focused on genetic engineering of traits directly related to traditional approaches for control of insects, weeds, and plant diseases. The short- to medium-term strategies of the farm-supply industries concentrate on market expansion or extension for registered crop chemicals through promoting herbicide tolerance and developing encapsulated embryos to replace natural seed. Longer-term corporate and university strategies assume some shift away from chemical-based crop protection toward biological pest controls and fertilizers (e.g., bioinsecticides, bioherbicides, biofungicides, and nitrogen fixation) (Fowler et al. 1988) and the development of genetically-engineered hybrids.

Significant improvements continue to be made in the methods available for identifying and isolating critical genes, introducing them into plants, and regulating their subsequent expression. Through DNA delivery techniques, such as *Agrobacterium*-mediated transformation and particle bombardment, important genes have been transferred to commercially valuable crops (Gasser and Fraley 1989). Transgenic plants have been produced in more than two dozen species, including herbaceous dicots such as tobacco, soybeans, cotton, carrots, tomatoes, potatoes, alfalfa, oilseed rape, and sunflower; woody dicots such as walnut and apple; and monocots such as rice, corn, and rye. Researchers at Monsanto predict that genetically engineered soybean, cotton, rice, corn, oilseed rape, sugarbeet, tomato, and alfalfa crops will enter the market place between 1993 and 2000 (Gasser and Fraley 1989).

Weed Control. Most of the research in the area of weed control centers on development of herbicide tolerance in crop species. Worldwide $10 billion are spent annually on herbicides. Of this, $3 billion are spent in the United States (Golden 1990). At least 27 corporations, including the world's eight largest pesticide companies, have research programs directed toward the development of herbicide-tolerant or herbicide-resistant crop varieties, with an expected market value exceeding $6 billion by the turn of the century (Goldburg et al. 1990). Industry researchers report that work has generally concentrated on those herbicides with properties such as high unit activity, low toxicity, low soil mobility, rapid biodegradation, and broad-spectrum activity against various weeds (Gasser and Fraley 1989).

Insect Resistance. Production of insect-resistant plants is another area of biotechnology applications with important implications for crop improvement. However, the number of research programs on insect resistance is only half that for herbicide tolerance, and the scope of work is surprisingly narrow. Some analysts estimate that 95 percent of the commercial research in bioinsecticides is concentrated on the introduction of *Bacillus thuringiensis* insect-control genes into crops. This bacterium produces insect-control proteins that are lethal to selected insect pests. Proteins from different strains of *B. thuringiensis* are toxic to lepidopteran pests (moths and butterflies), and in some cases to beetle and fly larvae, but are not toxic to beneficial insects (e.g., honeybees and ladybugs), other animals, or humans. Transgenic tomato, tobacco, and cotton plants containing a *B. thuringiensis* gene exhibited resistance to caterpillar pests in laboratory tests (Gasser and Fraley 1989). Since annual expenditures for insecticides against lepidoptera are estimated at $1.5 billion across all crops, commercial applications of these bioinsecticides may appear in the near future. However, recent evidence suggests that the more potent *B. thuringiensis* proteins may select for mutant insect pests and that insect resistance to *B. thuringiensis* protein is developing (Fowler et al. 1988). Consequently, other types of insecticidal molecules may be required to extend biotechnology approaches for insect control.

Disease Resistance. Development of disease-resistant plants is another important area of crop research. Significant resistance to tobacco mosaic virus (TMV) has been achieved by expressing the protein coat of TMV in transgenic plants. This approach has produced similar results in transgenic tomato, tobacco, and potato plants against a broad spectrum of plant viruses. For example, Monsanto researchers recently announced they had genetically engineered potatoes to resist the potato viruses PVX and PVY. There are no chemical controls for these viruses (Lawson et al. 1990). As a consequence, significant yield protection could be provided in important vegetable crops and in corn, wheat, rice, and soybeans. Moreover, limited success in engineering resistance to fungal diseases has been reported, although this research is in its early stages. For example, DNA Plant Technology has recently announced the successful field trial of plants genetically engineered to produce a high level of the antifungal enzyme, chitinase (*Genetic Engineering News* 1991). This, and similar work, is particularly important since fungi are the largest and most economically important group of plant pathogens, causing billions of dollars of crop losses and postharvest food spoilage each year (Timberlake and Marshall 1989). Current world spending on fungicides is estimated to be $2.3 billion.

Biofertilizers. In the area of biofertilizers, there is progress in increasing the nitrogen-fixing capacity of some leguminous plants. Biotechnica International has modified a bacterium (*Rhizobium meliloti*) that may increase the yield of infected alfalfa by 15 percent. If field tests are successful, the improved *R. meliloti* will be available as a soil inoculant in the early 1990s. According to a company representative, this genetically engineered bacterium could increase crop value as much as $1 billion (Sterling 1988). In another effort, modifying the expression of two specific genes in nitrogen-fixing symbionts resulted in a more than 10 percent increase in biomass production by the host plants under greenhouse conditions (Lindow et al. 1989). In other developments, scientists at three U.S. universities have successfully transferred nitrogen-fixing genes to non-nitrogen-fixing crops. The symbiotic associations among the crops and the various genes from nitrogen-fixing bacteria (e.g., *Rhizobium, Bradyrhizobium, Frankia*) are highly complex, but offer many important future possibilities. Scientists believe that ultimately it may be possible to manipulate both crop and rhizobial genes to achieve maximum efficiency of nodule formation and function and to tailor strains for unusual soil environments (Lindow et al. 1989, Sterling 1991). If transferring nitrogen-fixing genes to other crops proves to be commercially viable, the reduction in the use of nitrogen fertilizers could be substantial. However, nitrogen-fixation research may have the opposite effect. The development of plant varieties capable of utilizing synthetic fertilizers more effectively or absorbing greater quantities of them could lead to increased use of chemical fertilizers, thereby increasing production costs and environmental hazards.

Food Processing

The third major area of the food and agriculture system where biotechnological developments are already having an impact is food processing. Here techniques for the growth of and fermentation by yeast and bacteria, well-known to certain segments of the food-processing industry (e.g., cheese, bread-making, and alcohol production), will be equally important in the future. Genetic engineering is already being applied to bacteria, yeast, and other fungi to produce starter cultures with specific metabolic capabilities in food fermentation. Other biotechnology applications include plant tissue-culture technology for the production of plant-derived food ingredients; protein engineering to tailor enzymes with specific catalytic capabilities (e.g., thermostability, substrate specificity, or pH stability), so that these enzymes become more amenable to processing systems; and genetic engineering of microorganisms to enhance production of high-value food ingredients, facilitate waste management in the food industry, and improve food safety (Newell and Gordon 1986). These processes, combined with the new cell-culture techniques, are being used to transform the production of certain agricultural commodities into industrial processes. In principle, any commodity that is consumed in an undifferentiated or highly processed form could be produced in this manner, and product substitution could be easily introduced. Similarly, although with greater difficulty, tissue culture techniques could be used to produce edible plant parts *in vitro*. In short, agricultural production in the field could be supplanted by cell- and tissue-culture factories (Goodman et al. 1987).

Specific examples of current biotechnology applications in this area include: attempts to produce a natural vanilla product through phytoproduction in the laboratory, and genetic modification of oilseed crops to convert cheap oil (e.g., palm or soybean oil) into high-quality cocoa butter. Another example is the application of biotechnology to produce substitutes for sugar as an industrial sweetener. Major corporations in the United States and Europe, such as Unilever and Ingene, are attempting to use rDNA technology to produce the thaumatin protein, one of the sweetest known substances. The development of this product through genetic engineering may continue a transition to alternative sweeteners, reducing the market for beet and cane sugar and capturing the valuable sweetener market, currently worth $8 billion in the United States (Fowler et al. 1988, Rural Advancement Foundation International 1991).

Environmental Concerns

Accompanying these promising biological developments are increasing concerns about the environmental and ecological impacts of biotechnology. The perceived risks of biotechnology to the environment have become a major issue. Many scientists, environmentalists, and citizens' groups have raised safety issues regarding possible unexpected consequences of introducing new organisms, such as the production of a toxic secondary metabolite or protein toxin, or undesired self-perpetuation and spread of the organism. These organisms are viewed as inherently more unpredictable than chemical products, since living organisms will reproduce, grow, migrate, and mutate. Furthermore, while any given introduction has only a small chance of becoming a problem, each is characterized as a low-probability, high-incident risk, with long-term consequences that could be enormous and irreversible (Hoy 1991).

In agriculture some have suggested that living natural inputs may be even more dangerous to society than the artificial products they produce. Robert Goodman, the former Vice President of a leading agricultural biotechnology company, Calgene, and currently professor at the University of Wisconsin, notes that some of nature's allopathic chemicals include arsenic and cyanide. "Just because it is natural doesn't mean it is safe" (Quoted in Fowler et al. 1988:85). Critics of the research, however, believe that the release of genetically engineered organisms may result in some unintended and perhaps permanent damage or loss to the environment. The analogy is often made to previous problematic introductions of exotic organisms into North America (e.g., kudzu vine, African bee, gypsy moth, and zebra mussel). In addition, there is concern that genetically engineered crops bearing resistance to major herbicides may become weeds or cross with weedy relatives and spread resistance into sectors of the weed flora (Klassen 1989). Furthermore, critics warn, engineered organisms might also become pests, displacing existing plants and animals, disrupting the functioning of ecosystems, reducing biological diversity, altering the composition of species, threatening the extinction of various species, or even changing climate patterns (Mellon 1987, Gould 1988).

These critics have successfully sought court injunctions to suspend work and to prevent research that entails the release of genetically engineered microbes or other organisms in a natural setting. Extensive governmental mechanisms for review and approval of biotechnological research have been developed in an effort to address some of these important concerns. Essential in this process is more knowledge of the factors that affect the abilities of species to survive, proliferate, and disperse in nature and a greater understanding of the potential for genetic exchange between species.

Government and University Biotechnology Investment

The knowledge and tools generated by molecular biology and biotechnology have stimulated a great deal of enthusiasm and led to the investment of large sums of money in an effort to pursue knowledge in this area. At the federal level, financial support for biotechnology has grown steadily since the mid-1980s, reaching $3.8 billion in 1991. The President's budget for 1992 calls for an 11 percent increase, to $4.1 billion. While 80 percent of the federal budget for biotechnology has been devoted to National Institutes of Health programs, support for agricultural biotechnology has been relatively meager, constituting less than 3 percent of federal expenditures for biotechnology (Metheny and Monahan 1991). There is, however, optimism that over the next five years agricultural biotechnology will receive substantial increases at the federal level through the U.S. Department of Agriculture's National Research Initiatives Program for Research on Agriculture, Food and Environment (U.S. Department of Agriculture 1991).

In the United States, universities have been major contributors to the development of biotechnology. Industry has looked to universities as centers of cutting-edge research and collaborative work, while states have viewed their universities as the basis of their biotechnology efforts and as a means of attracting biotechnology companies. Of the 33 states reporting state-supported biotechnology programs (total annual funding of approximately $150 million), 28 primarily relied on higher-education institutions for the design and performance of biotechnology research. Moreover, in 14 states, universities have been the impetus for creating biotechnology programs (Office of Technology Assessment 1988).

Governments of several other nations are also investing in biotechnology. The support is substantial, but is considerably less than that provided by the U.S. government. Unlike the United States' emphasis on basic research, the Japanese and many European governments have focused on the commercial goals of biotechnology and have supported applied research and development efforts that emphasize the transfer of research from government laboratories to industry. Japan's Ministry of International Trade and Industry (MITI), which guided that nation's development efforts in automobiles and computers, has designated biotechnology as a strategic industry and a national priority. MITI established groups of companies and provided seed money in four areas of biotechnology: gene-splicing, large-scale cultivation of cells, bioreactors (microbes that change one chemical into another), and bioelectronics (Wysocki 1987). Research and development support to Japanese firms has been boosted to several hundred million dollars; in 1986, the government's biotechnology budget was $196 million. Most recently Japan's Ministry of Agriculture, Forestry and Fisheries announced a seven-year, $25-million project to produce a high-resolution map of the rice genome and sequence its most important genes (Crawford 1991). This strong Japanese emphasis on commercial development of biotechnology is predicted to account for more than 10 percent of the Japanese gross national product by the year 2000 (Ward 1989).

Of the European nations, the United Kingdom is the most involved in biotechnology. In 1989, the United Kingdom allotted $100 million of public funds for biotechnology-related research to strengthen its initial investments in this area. An additional $25 million will be provided to the Agricultural and Food Research Council for plant molecular biology (Newmark 1989). Four other European nations (Spain, the Netherlands, France, and West Germany) also have national programs to support and coordinate biotechnology efforts. In West Germany during the mid-1980s, three federally sponsored biotechnology research institutes joined with industry to invest an estimated $280 million in biotechnology research (Office of Technology Assessment 1984). In addition to the efforts of individual nations, the Commission of European Communities (CEC) budgeted approximately $150 million for biotechnology research for 1987-91.

Commercial Agricultural Biotechnology

The emergence of a worldwide biotechnology industry parallels these rapid scientific developments and government and university investments. This industry began in the United States in 1976 with the founding of Genentech (the first firm to exploit rDNA technology) and has grown rapidly. By 1984, 219 U.S. companies were pursuing applications of biotechnology (Office of Technology Assessment 1984). Just four years later OTA (1988) identified 403 dedicated biotechnology companies. The most recent survey by *Genetic Engineering News* (1990) included 751 companies worldwide, with 592 in the United States.

While most companies are developing applications in pharmaceuticals and diagnostics, a large number are involved in agricultural biotechnology. For example, in 1984 28, percent of the U.S. companies were involved in animal agriculture, while 24 percent were involved in plant agriculture (Office of Technology Assessment 1984) Worldwide, in 1990 242 companies (32%) indicated an emphasis on agricultural-related products and processes (*Genetic Engineering News* 1990).

Despite this growth, consolidation within the industry and the predominance of a small number of large corporations have begun to slow the formation of new firms. Indeed 80 percent of the venture capital has been invested in just ten companies; health-care applications account for 75 percent of all investments; agriculture accounts for only 16 percent. Nearly all of these biotechnology firms lose money. The agricultural biotechnology companies, like their counterparts in pharmaceuticals and chemicals, have recognized the need to merge and consolidate among themselves and to establish strategic relationships with the large agrichemical and seed corporations and multinational food processors. These agricultural venture capital firms have found it increasingly difficult to raise additional capital, have encountered regulatory delays in conducting research and bureaucratic delays in obtaining patents, and have generally lacked the complete set of capabilities required to market a new product (i.e., research development, formulation, scale-up, clinical and field testing, manufacturing, and marketing). In a recent survey, biotechnology-firm executives predicted that within 10 years roughly half of all U.S. biotechnology firms would fail, merge, or form cost-sharing alliances (Naj 1989).

By the early 1980s, large multinational corporations began to recognize the potential of biotechnology, to develop their own research and development capacities, and to establish strategic partnerships with small biotechnology firms. These corporations have diversified into every field or specialty that uses living organisms as a means of production. The new biotechnologies appear to be further reducing distinctions among traditional industrial sectors, virtually eliminating corporate boundaries. For example, large nonpharmaceutical corporations, such as DuPont, Nestle, Monsanto, Procter and Gamble, and Dow Chemical, have developed a new emphasis on and invested heavily in pharmaceutical efforts through biotechnology research and development. Moreover, large multinational corporations specializing in oil, chemicals, food, and pharmaceuticals (including American Cyanamid, Campbell's Soup, Ciba-Geigy, DuPont, Eli Lilly, Hershey Foods, Lubrizol, Monsanto, Rhône-Poulenc, RJR Nabisco, Rohm and Haas, Shell, Sandoz, Standard Oil, and W. R. Grace) have taken the leadership in agricultural biotechnology in the latter half of the 1980s.

Today, a wide range of alliances have been formed between the 70 major U.S. corporations investing in biotechnology and the venture-capital firms. While collaborations are not always between large corporations and small ones, the Office of Technology Assessment estimates that three-quarters of the more than 1,000 joint agreements are of this nature. About a fifth of the major U.S. corporations are committed to agriculture. Surveys of these agriculture-related firms reveal employment of more than 1,000 molecular biologists and annual investments in agricultural biotechnology research and development of over $200 million. Most of the agricultural biotechnology work continues to be conducted by U.S. firms, followed by European, Latin American, Canadian, and Japanese firms (Office of Technology Assessment 1988).

For the year 2000, estimates of biotechnology's market impact on agriculture range from $12.6 to $67 billion (Fowler et al. 1988). According to some analysts, after the turn of the century, approximately $12 billion of the estimated $30 billion world commercial seed market will contain contributions from biotechnology (Van Brunt and Klausner 1987). Furthermore, packages, including a specific pesticide or herbicide and plants resistant to that chemical, could constitute a sizable portion of the $10 billion agricultural chemical market (Office of Technology Assessment 1988). A recent analysis of world agricultural markets (Ratafia and Purinton 1988) predicted rapid commercialization for some genetically manipulated crops, including tomatoes, potatoes, certain other vegetables, and sugar cane. In the area of food processing, a Japanese industry evaluation estimated the economic impact of biotechnology to be $17.2 billion in 2000, well above their $12.8 billion estimate for pharmaceuticals (Newell and Gordon 1986). Others, not as optimistic, predict annual real growth will be a moderate 5 percent annually (Office of Technology Assessment 1988).

These biological developments and the emergence of commercial agricultural-biotechnology firms have occurred at a time of major restructuring of the agricultural sector, characterized by concentration of farm ownership and of farm-input and food-processing industries and by increasing industrialization of food production. In the farm sector, an accelerating concentration has reduced the number and increased the size of farms. From 1915 to 1945 the number of farms in the United States decreased about 8 percent, while from 1945 to 1973 it decreased 55 percent, from 5.9 million to 2.8 million. By the late 1980s, the number was below 2 million. Market shares have also become increasingly concentrated. By the year 2000, 50,000 American farms are expected to account for at least 75 percent of the nation's agricultural output (Klausner 1986). In addition, in 1979, 15 companies provided 60 percent of all farm inputs, 49 companies processed 68 percent of all the food, and 44 companies received 77 percent of all wholesale and retail food revenues (Vogeler 1981).

Concentration is even greater in specific industries. By the 1970s, four tractor companies accounted for 83 percent of all tractor sales (Vogeler 1981). In the mid-1970s, 30 U.S. manufacturers were engaged in pesticide development. Only 12 were doing so by the late 1980s, and of these only 6 are expected to continue until the turn of the century. The pesticide industry is dominated worldwide by seven transnationals (Bayer, Ciba-Geigy, DuPont, Hoechst, Imperial Chemical Industries, Monsanto, and Rhône-Poulenc), each with sales in excess of $1 billion dollars and together capturing 63 percent of the global market (Fowler et al. 1988).

This concentration, accompanied by horizontal and vertical integration across industrial sectors, reflects the mergers and acquisitions in the agricultural-input and food-processing industries, as many traditionally nonfood companies significantly expand their investments. Most of the major seed companies have been purchased by large multi-national chemical, oil, and pharmaceutical corporations. Of the seven transnationals dominating the pesticide industry, five also rank among the world's twenty largest seed companies; only Bayer and DuPont have marginal seed interests. Moreover, of the ten largest seed companies, eight (Ciba-Geigy, Imperial Chemical Industries, Lafarge, Pfizer, Sandoz, Shell, UpJohn, and Volvo) also have significant interests in crop chemicals. Similarly, the food-processing industry is becoming increasingly concentrated. In 1987, 100 U.S. food processors accounted for approximately 80 percent of all sales in the industry. In fact, ten companies controlled nearly one-third of the market, up 30 percent from 1982. Mergers in Europe and the United States have also dominated this industry. More than one-quarter of the 100 processors doing business in 1982 have been purchased or merged, with several major acquisitions involving the tobacco industry (e.g., R.J. Reynolds acquired Nabisco, Philip Morris acquired General Foods, Fowler et al. 1988). Most analysts predict that this pattern, in conjunction with increasing concentration and consolidation in the industry, will continue in the future. By the next century, most of the food system could be controlled by a small number of highly diversified, multinational corporations.

Impacts of Biotechnology

The impacts and consequences of biotechnology on science, the food and fiber-system, and society are just emerging. Consequently, the predicted implications of biotechnology can represent only potential scenarios. The effects may include impacts on science, farmers and rural communities, consumers, the structures of agribusinesses and industries, global markets, and developing countries.

Impact on Science

Perhaps the most dramatic immediate impact of the new biotechnologies has been on science itself. While some argue that biotechnology is just a continuation of the application of biological techniques to improve plants, animals, and microorganisms, molecular biologists contend that biotechnology has revolutionized biology.

The techniques and tools of biotechnology are facilitating efforts to understand the intricate and complex functioning of living organisms at the molecular and cellular levels and are accelerating the accumulation of knowledge in traditional disciplines, such as biology, genetics, plant physiology, and biochemistry. Moreover, biotechnology, particularly in agriculture, may truncate both the time and space required to develop new plant, animal, and food products. Finally, it may complement and extend traditional methods used to enhance agricultural productivity and nutritional quality.

Still, several negative impacts of biotechnology on science may exist. Biotechnology and molecular biology continue and extend the basic methods and approaches of modern science. Their perspective, often called "logical positivism," attempts to reduce nature to small, definable pieces, subject to human manipulation, and separated from broader questions of value. From this perspective, scientists control, measure, reduce, and divide nature in order to generate knowledge. One concern is that this approach, while providing only partial knowledge, has become the dominant epistemology, often to the exclusion of important alternative ways of knowing. As a consequence, whole-plant- and whole-animal-level research (such as traditional plant breeding) and systems-level research programs (such as agroecology, farming systems, and social assessments), important complements to a comprehensive biotechnology research agenda, lack adequate support.

In U.S. colleges of agriculture, for example, employment opportunities in agricultural biotechnology and molecular genetics have grown significantly [260% increase between 1982 and 1988, National Association of State Universities and Land-Grant Colleges (NASULGC) 1989], while employment in plant and animal breeding has declined. In fact, many of the positions in molecular biology were created by reducing the scope of conventional breeding programs (Busch et al. 1991).

Another impact of biotechnology on science is the increased concentration of research funds and scientific talent at a small number of public and private institutions. For example, in the public sector, every U.S. state could afford and has had a conventional breeding program. Every state cannot afford and will not be able to have a comprehensive agricultural-biotechnology program. Currently, eight states account for more than half of the state experiment station expenditures and nearly half of all the scientists' years devoted to biotechnology research (NASULGC 1989).

The new biotechnologies are also contributing to the changing division of labor between universities and industries, with a number of concomitant impacts. Partnerships between universities and industries have existed for several decades, but the new types of university and industry relationships in biotechnology (e.g., centers, institutes, research parks, public corporations) are more varied, more aggressive, and more experimental. The legal/contractual bases for these relationships include: large grants and contracts between companies and universities in exchange for patent rights and exclusive licenses to discoveries; programs and centers, organized with industrial funds at major universities, that give participating private firms privileged access to university resources and a role in shaping research agendas; professors, particularly in the biomedical sciences, serving in extensive consulting capacities on scientific advisory boards or in managerial positions of biotechnology firms; and faculty receiving research funding from private corporations in which they hold significant equity (Lacy and Busch 1989). In a recent survey of biotechnology researchers at 40 major universities, 47 percent of these scientists consulted with outside corporations and 8 percent held equity in firms providing products or services directly related to their own university research (Blumenthal et al. 1986).

The consequences of these collaborations may be both positive and negative. First, university and industry collaboration may bring useful products to market more rapidly and may promote U.S. technological leadership in a changing world economy. This is a major motivation behind a number of recent funding policy statements and laws of the federal government, requiring such collaboration to become eligible for federally funded research. Second, in light of funding stagnation at the federal and the state levels, such collaborations are a means of raising new funds for university research. Third, these joint efforts may expand the scientific network, increasing communication between some industry and university scientists (Lacy and Busch 1992).

A number of concerns have been voiced regarding these new relationships, however. First, long-term research, previously a major emphasis of the public sector, may decline. The private sector has short-term proprietary goals; as a consequence, funding for research is also generally short-term. In contrast, nearly all of the NIH extramurally funded programs and the USDA Hatch formula-funded projects are for three years or longer. Moreover, the nature of the work may change. In a recent study of agricultural college biotechnology researchers, Curry and Kenney (1990) note that measures of scholarly and academic productivity declined as the percentage of the scientists' research budgets provided by industry increased.

A second issue is the potential restriction of communication. Proprietary agendas have begun to inhibit the free flow of information among biotechnology scientists and have raised concern about access to information. This is particularly true for university scientists with private-sector grants, who often must delay public discussion of their work, or its results, pending review by the sponsoring company. In one study 25 percent of the industry-sponsored biotechnology faculty reported that results of their research belong to the sponsoring firm and cannot be published without that firm's consent. Forty percent reported that their collaboration resulted in unreasonable delays in publishing (Blumenthal et al. 1986).

Even some scientists with public funding feel inhibited about discussing their work, for fear that a private company with the personnel, money, equipment, and/or time will exploit their ideas and perform the experimental work before they can. The net effect of these various developments appears to be a reduction of the free flow of information. Open communication is considered fundamental in public-sector science, and indeed, one industry scientist, in a 1987 personal interview, remarked that more knowledge is generated in an open environment for scientists (Lacy and Busch 1989).

A third concern is the potential for conflict of interest and/or scientific misconduct. In interviews, public- and private-sector scientists alike stressed the potentially detrimental effects of granting private patents for work done in the public sector. These effects include favoritism, unwarranted financial advantage through privileged use of information or technology derived from publicly funded research, constraints on sharing of germplasm, and shelving of research of interest to the public but not to the corporation (Lacy et al. 1988).

Recently, Derek Bok in his final President's report to the Harvard University Board of Overseers, warned that the commercialization of universities may be the most severe threat facing higher education. Mr. Bok noted that as universities become "more entrepreneurial they appear less and less as charitable institutions seeking truth and serving students and more and more as huge commercial operations that differ from corporations only because there are no shareholders and no dividends." He concluded by saying that "it will take very strong leadership to keep the profit motive from gradually eroding the values on which the welfare and reputation of universities ultimately depend" (McMillen 1991:A31).

Impact on Farmers and Rural Communities

The impact of agricultural change on rural communities is largely proportional to the level of local dependence on agriculture. Today, nationwide, fewer than 40 congressional districts have more than 20 percent of their population living on farms (Sundquist and Molnar 1991). The overwhelming majority of the farms that once existed in the United States no longer exist, and production is highly concentrated among the remaining farms. The largest 13 percent of farms now produce more than 75 percent of the value of total production. Many small farms are buffered from the effects of technological change, since for the owners, the farm is no longer their primary source of income. Consequently, biotechnology will probably have less impact on the total number of farms than the mechanical and chemical technologies adopted by farmers during the last 50 years (Buttel 1989). Moreover, it is likely that biotechnology will not greatly accelerate the decline in the number of farms, although it will certainly maintain present trends.

The Bureau of Labor Statistics predicts that farming will continue to be one of the fastest-declining occupations in the next decade (an estimated 28% decrease between 1990 and 2000).

The extent of biotechnology's influence on the trend towards fewer and larger farms depends, in part, on how adoption affects the cost structure of farming. If biotechnology developments significantly alter costs, returns, competitive positions, and the benefits of location of production, and if certain trade and farm policies are implemented, the potential impact of biotechnology could be relatively important. The Office of Technology Assessment (1986) has argued that these new technologies will be adopted by well-financed, innovative farmers who are presumed to run the comparatively large farms. However, others have argued that biotechnology innovations will provide widespread benefits to the full range of farmers because new technologies will be used in traditional ways. Regardless of which farmers are likely to benefit, however, biotechnology will probably increase the value-added off-farm at the expense of value-added on-farm (Goodman, Sorj, and Wilkinson 1987).

Other significant changes in the farming community may result if the information and products of this technology bypass the USDA Extension Service and the agricultural cooperatives. Previously, products and information from biological research have been disseminated through the Extension Service. However, if the development of new seed-chemical packages, animal-diagnostic kits, vaccines, and growth promotants through biotechnology arise from private research, public-sector scientists will have limited knowledge with which to support Extension programs. As a consequence, the Extension Service, and (potentially) agricultural cooperatives, might be reduced to playing a secondary role in farm change. Moreover, many agriculturally based rural communities may continue the ongoing process of shrinkage and consolidation, as producers, local-supply companies, and local-marketing firms decline in numbers (Lacy et al. 1990).

Biotechnology may also accelerate the trend toward contract integration, already a common practice in the United States for such commodities as poultry and most processed vegetables. Currently, contracts for vegetables specify the seeds and chemicals to be used, set planting and harvesting schedules, and prescribe other aspects of production as desired by the vegetable processor. These arrangements further reduce the autonomy of farmers, as well as their contact with and need for Extension services, agricultural cooperatives, and local farm suppliers. The new biotechnologies may also restructure the relationship between farmers and researchers. Until very recently, farmers were seen as the primary clientele of public-sector research. However, the new university-industry partnerships in biotechnology and agricultural research have increasingly resulted in an amplification of the demands of the agribusiness sector. As a result, it is quite possible that only problems of interest to this group will be on public-research agendas (Goodman et al. 1987).

Impact on Consumers

The new biotechnologies expand and extend researchers' abilities to improve plants, animals, and microorganisms. For consumers this could mean dramatic improvements in the productivity and efficiency of food production and processing, and the expansion and extension of food and nonfood uses of raw agricultural commodities. Consumers could benefit in the form of reduced prices, increased food safety, and more-nutritious foods. The new technologies also have the potential to change the very nature of food itself. With molecular-biology approaches, scientists can move genetic material among plants and between plants, animals, and microorganisms. It is now possible to consider the production of new fabricated foods; basic foods could be broken down into their component parts (e.g., starch, fat, and sugar) and recombined into wholly new types of food. Such new forms of food may be undesirable to consumers, as it may become far more difficult to determine food composition and to maintain a balanced diet.

Public concern about recent scientific developments, including biotechnology, has led to a decline in public confidence in science and to serious considerations of the limits of science. Biotechnology has stimulated new moral and ethical debates regarding a wide range of concerns about science. These concerns extend beyond human health, environmental risk, food safety, and animal health, and include such issues as socioeconomic consequences and the morality of tampering with nature and life itself (Lacy et al. 1991). The impacts these consumer concerns may have on the development of biotechnology is contingent in substantial measure on the degree of consumer mobilization around these concerns by key consumer and environmental groups and on the nature of their actions (e.g. boycotts, product labeling, legal action).

Impact on Agribusiness

During the 1980s, John Hardinger, Director of Biotechnology at DuPont's Agricultural Products department, viewed biotechnology as a force to not only restructure farming but also catalyze major change in the structure of worldwide agribusiness. He noted that the application of molecular biology permits logical linkage of the various segments of the world's largest industrial sector with other economic sectors that was never before practical. This $1.3 trillion agribusiness sector (not counting feed and fiber) consists of four basic elements: input suppliers, growers, processors, and consumers. This system has experienced mechanical and chemical eras which contributed to increased productivity and efficiency. According to Hardinger, the new biological and biotechnology era will further increase efficiency and productivity, as well as provide the ability to improve the quality of food and feed. Furthermore, it will lead to consolidation and new forms of vertical integration of the food industry (Busch et al. 1991).

As discussed in the section on Commercial Agricultural Biotechnology, this concentration, accompanied by horizontal and vertical integration across industrial sectors, has been occurring in the agribusiness sector in the last decade. Mergers, acquisitions, and concentration in the agricultural-input and food-processing industries have occurred, as traditionally nonfood industries such as chemicals, petroleum, pharmaceuticals, and tobacco expand their investments. Most analysts predict biotechnology will continue and accelerate this trend towards increasing concentration of control by a small number of large multinational corporations.

Consequently, development and commercial control will be in the hands of corporations that transcend geographic boundaries and hold limited national allegiance. Within this context, people question how we can ensure democratic participation in the decision-making processes surrounding the development and commercialization of biotechnology. This is difficult within national boundaries and generally prohibitive internationally, given current governmental structures.

International Impacts

For some, the new technologies offer the hope of increasing crop yields in nations where population growth is outstripping the food supply. It has been proposed that the direct use of molecular biology, in conjunction with plant propagation and breeding, could dramatically increase crop productivity and overall food production in developing countries. Tissue-culture techniques are already creating more drought- and disease-resistant varieties of cassava, oil palm, and groundnuts. Embryo transfer may increase the reproductive capacity of livestock. Genetically engineered vaccines and drugs may control or cure fatal or debilitating diseases such as onchoceriasis (river blindness), guinea worm disease, schistosomiasis, and trypanosomiasis, thus opening up new agricultural and grazing areas in Africa (Gibbons 1990, Barker and Plucknett 1991).

Yet despite biotechnology's great promise for feeding the world's rapidly growing population, particularly in developing nations, policy makers admit that it will be difficult to ensure that this technology has the desired positive effects. First, there is legitimate concern that the developed nations will use their technology to undercut traditional Third World exports, such as sugar, vanilla, cocoa, and other important cash crops. Even moderate success in realizing crop relocation, laboratory-based phytoproduction, or product substitution would have profound effects around the world, the most immediate and most important being the restructuring of global markets (Persley 1990).

Environmental risk is another important issue. Because even planned release of genetically modified organisms may inadvertently have hazardous effects on the ecosystem, the public in many industrialized countries has pressed for the adoption of more stringent safety regulations. These regulations, however, may restrict biotechnology experiments. As a consequence, some researchers and biotechnology companies may relocate their experiments to countries with limited or no safety regulations. Thus, potentially hazardous biotechnology experiments may be conducted in Third World countries.

A further concern is that biotechnology will increase the disparities between the developed and developing nations. With the shift of applied research and associated product development from the public to the private sector, benefits of the new biotechnologies may become less widely available. Moreover, the products developed are not likely to be ones that are important to developing countries, particularly those in the tropics. Biotechnology research has emphasized temperate-zone animal reproduction, breeding, veterinary health care, and nutrition, and temperate-zone plant improvement. Little or no work is currently being directed at improvement of tropical crops important to developing countries. This could further widen the gap between the agricultural production methods of the developed versus the developing countries (Deo et al. 1989).

In conclusion, agricultural biotechnology may shift the geographic location of agricultural production from one Third World country to another, or from the Third World to the First World. For many Third World countries dependent on one or two agricultural commodities for their continued viability, such production and market restructuring could disrupt or collapse their present markets. Significant numbers of farmers and farm workers might find themselves with no market for their products. A decline in export income could increase the already high Third World debt and exacerbate the deficit in balance of payments in Third World countries. Political instability, already a problem in the developing world, would doubtless increase.

Changes in world trade patterns are also likely to affect First World nations. Developing countries experiencing economic deterioration would no longer serve as a main market for developed country exports, possibly creating economic and social stress in the developed nations as well. For the continued well-being of an increasingly global economy, a conscientious effort to help developing nations acquire the appropriate technology, establish and maintain a supporting infrastructure for applied research, and improve their capacity to evaluate new technology in terms of their own public good becomes essential.

Conclusions

This chapter has outlined public perceptions of the new biotechnologies, certain scientific developments in biotechnology, the roles of governments, universities, and industries in its development, and the broad technical, social, and economic issues and impacts associated with its emergence. Despite the relative infancy of biotechnology, many scientists, government officials, industry analysts, and the public envision a multibillion dollar market for its products in the foreseeable future. These potential impacts of biotechnology, however, raise complex ethical and policy questions. These new technologies present new social choices and introduce far more options than we can explore. Biotechnology now permits the design of future agricultural systems without providing any guidance for the directions of technology and systems development and the assessment of embedded social choices. To address these important issues a more careful review and monitoring of scientific developments are essential, as is a detailed assessment of the various impacts of technological change.

The chapters that follow analyze the development of a particular technological change in the dairy industry. A careful examination of the development and use of genetically engineered bovine somatotropin should help to clarify the issues surrounding this highly controversial technology. Furthermore, it should provide insights for analyzing future biotechnologies and for eventually developing the framework for effective *ex ante* evaluation of emerging technologies. Finally, assessments, such as this one, need to be incorporated into our decision-making and into the development of policies and long-range planning capacities so that we can effectively address future scenarios. Whose needs and goals will be served and whose will be neglected by the new biotechnologies are among the most important agricultural and social questions of the coming decade.

Acknowledgments

This chapter is based in part on material appearing in L. Busch, W. B. Lacy, J. Burkhardt, and L. R. Lacy. 1991. *Plants, Power, and Profit: Social, Economic, and Ethical Consequences of the New Biotechnologies.* Cambridge, MA: Basil Blackwell. We appreciate the helpful review and suggestions of Dr. F. H. Buttel on an earlier draft of this chapter.

References

Associated Press. 1989. "Foreign gene injected into human for first time." *Lexington Herald-Leader*, 23 May, A3.

Barker, R., and D. Plucknett. 1991. "Agricultural biotechnology: A global perspective." Pp 107-120 in B. R. Baumgardt and M. A. Martin (eds). *Agricultural Biotechnology: Issues and Choices.* West Lafayette, IN: Purdue University.

Blumenthal, D. M., M. Gluck, K. S. Louis, M. A. Stoto, and D. Wise. 1986. "University-industry research relations in biotechnology: Implications for the university." *Science* 232:1361-6.

Busch, L., W. B. Lacy, J. Burkhardt, and L. R. Lacy. 1991. *Plants, Power and Profit: Social, Economic, and Ethical Consequences of the New Biotechnologies*. Cambridge, MA and Oxford, UK: Basil Blackwell.

Buttel, F. H. 1989. "How epoch making are the new technologies? The case of biotechnology." *Sociological Forum* 4(2):247-261.

Crawford, R. 1991. "Gene mapping Japan's number one crop." *Science* 252:1611.

Curry, J., and M. Kenney. 1990. "Land grant university-industry relationships in biotechnology: A comparison with the non-LGU research universities." *Rural Sociology* 55(1):44-57.

Deo, S. D., J. S. Silva, L. Busch, W. Lacy, and N. Mohseni. 1989. "Agricultural biotechnology in India and Brazil: Creating new technological dependencies." *International Journal of Contemporary Sociology* 26(3-4):217-233.

Dixon, B. 1989. "Biotech's major impact yet a decade away." *Bio/Technology* 7:420.

Fowler, C., E. Lachkovics, P. Mooney, and H. Shand. 1988. "The laws of life: Another development and the new biotechnologies." *Development Dialogue* 1-2:1-350.

Gasser, C. S., and R. T. Fraley. 1989. "Genetically engineering plants for crop improvement." *Science* 244:1293-9.

Genetic Engineering News. 1990. "1991 GEN guide to biotechnology companies." *Genetic Engineering News* 10(10):17-53.

_____. 1991. "DNAP announces successful trial of altered plants." *Genetic Engineering News* 11(6):19.

Gibbons, A. 1990. "Biotechnology takes root in the Third World." *Science* 248: 962-963.

Goldburg, R., J. Rissler, H. Shand, and C. Hassebrook. 1990. *Biotechnology's Bitter Harvest: Herbicide-Tolerant Crops and the Threat to Sustainable Agriculture*. The Biotechnology Working Group.

Golden, M. 1990. "Novo Vordisk establishes R&D biopesticides facility in U.S." *Genetic Engineering News* 10(11):1,15,16.

Goodman, D., B. Sorj, and J. Wilkinson. 1987. *From Farming to Biotechnology: A Theory of Agro-Industrial Development*. Oxford, UK: Basil Blackwell.

Gould, F. 1988. "Evolutionary biology and genetically engineered crops." *Bioscience* 38:26-33.

Hilts, P. J. 1991. "Gene-altered pigs produce key part of human blood." *New York Times* June 16:1,16.

Hoban, T. 1990. "Public attitudes toward bovine somatotropin." Paper presented at 39th Annual Dairy Conference.

Holden, C. 1990. "Drugs and biotechnology." *Science* 248:964.

Hoy, M. A. 1991. "Biotechnology and the environment." *Science* 253:89-90.

Klassen, W. 1989. "Public concerns regarding applications of biotechnologies." *Biodeterioration Research* 2:7-17.

Klausner, A. 1986. "Gains, hardships to stem from agbiotech." *Bio/Technology* 4:385.

Lacy, W., L. Lacy, and L. Busch. 1988. "Agricultural biotechnology research: Practices, consequences, and policy recommendations." *Agriculture and Human Values* 5 (3):3-14.

Lacy, W. B., and L. Busch. 1989. "Changing division of labor between the university and industry. The case of agricultural biotechnology." Pp 21-50 in J. Molnar and H. Kinnucan (eds). *Biotechnology and the New Agricultural Revolution*. American Association for the Advancement of Science Symposium Series. Boulder, CO: Westview Press.

____. 1992. "The fourth criterion: Social and economic impacts of agricultural biotechnology." NABC Report #3, Agricultural Biotechnology at the Crossroads. Ithaca, NY (forthcoming).

Lacy, W. B., L. Busch, and W. D. Cole. 1990. "Biotechnology and agricultural cooperatives: Opportunities, challenges, and strategies for the future." Pp 69-82 in D. J. Webber (ed). *Biotechnology: Assessing Social Impacts and Policy Implications*. New York: Greenwood Press.

Lacy, W., L. Busch, and L. Lacy. 1991. "Public perceptions of agricultural biotechnology." Pp 139-166 in B. R. Baumgardt and M. A. Martin (eds). *Agricultural Biotechnology: Issues and Choices*. West Lafayette, IN: Purdue University.

Lawson, C., W. Kaniewski, L. Haley, R. Rozman, C. Newell, P. Sanders, and N. Tumer. 1990. "Engineering resistance to mixed virus infection in a commercial potato cultivar: Resistance to potato virus X and potato virus Y in transgenic Russet Burbank." *Bio/Technology* 8(2):127-43.

Lindow, S. E., N. J. Panopoulous, and B. L. McFarland. 1989. "Genetic engineering of bacteria from managed and natural habitats." *Science* 244:1300-7.

Martin, M. A., and B. R. Baumgardt. 1991. "The origins of biotechnology and its potential for agriculture." Pp 2-21 in B. R. Baumgardt and M. A. Martin (eds). *Agricultural Biotechnology: Issues and Choices*. West Lafayette, IN: Purdue University.

McMillen, L. 1991. "Quest for profit may damage basic values of universities, Harvard's Bok Warns." *The Chronicle of Higher Education* XXXVII(32):A1,A31.

Mellon, M. 1987. *Biotechnology and the Environment: A Primer on the Environmental Implications of Genetic Engineering*. Washington, D.C: National Wildlife Federation.

Metheny, B., and J. Monahan. 1991. "President's FY92 budget proposes $4.1 billion for biotechnology research and development." *Genetic Engineering News* 11(3):3.

Monsanto. 1989. "Farming: A picture of the future." *Science* 240:1384.

Naj, A. K. 1989. "Clouds gather over the biotech industry." *Wall Street Journal*, 30(1) B1,B5.

National Association of State Universities and Land-Grant Colleges (NASULGC). 1989. *Emerging Biotechnologies in Agriculture: Issues and Policies*. Progress Report VIII. Washington, DC: NASULGC, Division of Agriculture, Committee on Biotechnology.

Newell, N., and S. Gordon. 1986. "Profit opportunities in biotechnology for the food processing industry." Pp 297-311 in S. K. Harlander and P. Labuze (eds). *Biotechnology in Food Processing*. Park Ridge, NJ: Noyes Publications.

Newmark, P. 1989. "UK biotech spending spree of a sort." *Bio/Technology* 7:334.

Office of Technology Assessment. 1984. *Commercial Biotechnology: An International Analysis*. OTA-BA-218. Washington, DC: U.S. Government Printing Office.

____. 1986. *Technology, Public Policy and the Changing Structure of American Agriculture*. OTA-F-285. Washington, DC: U.S. Government Printing Office.

____. 1987. *New Developments in Biotechnology: Public Perceptions of Biotechnology*. OTA-BP-BA-45. Washington, DC: U.S. Government Printing Office.

____. 1988. *New Developments in Biotechnology: U.S. Investments in Biotechnology*. OTA-BA-360. Washington, DC: U.S. Government Printing Office.

____. 1991. *U.S. Dairy Industry at a Crossroad: Biotechnology and Policy Choices*. OTA-F-470. Washington, DC: U.S. Government Printing Office.

Persley, G. J. 1990. *Beyond Mendel's Garden: Biotechnology in the Service of World Agriculture*. Biotechnology in Agriculture Series, No. 1. Oxon, UK: CAB International.

Pursel, V. G., C. A. Pinkert, K. F. Miller, D. J. Bolt, R. G. Campbell, R. R. Palmiter, R. L. Brinster, and R. E. Hammer. 1989. "Genetic engineering of livestock." *Science* 244:1281-7.

Ratafia, M., and T. Purinton. 1988. "World agricultural markets." *Bio/Technology* 6:280-1.

Rural Advancement Foundation International. 1991. "Update: Vanilla and biotechnology; Genetically engineered oilseed plants." *RAFI Communique*. July: 1-4.

Sterling, J. 1988. "Agbio products edge closer to market place." *Genetic Engineering News* 8(5):1,15,16.

____. 1991. "Outside the ivy halls." *Genetic Engineering News* (6):8.

Sun, M. 1988. "Designing food by engineering animals." *Science* 240:240.

Sundquist, W. B., and J. J. Molnar. 1991. "Emerging biotechnologies: Impact on producers, related businesses, and rural communities." Pp 23-40 in B. R. Baumgardt and M. A. Martin (eds). *Agricultural Biotechnology: Issues and Choices*. West Lafayette, IN: Purdue University.

Timberlake, W. E., and M. A. Marshall. 1989. "Genetic engineering of filamentous fungi." *Science* 244:1313-25.

U.S. Department of Agriculture. 1991. *1992 Program Plan for the National Initiative for Research in Agriculture, Food and Environment*. Washington, DC: USDA, Office of the Assistant Secretary, Science and Education.

Van Brunt, J., and A. Klausner. 1987. "The Bio/Technology roundtable on plant biotech." *Bio/Technology* 5:128,130-3.

Vogeler, I. 1981. *The Myth of the Family Farm: Agribusiness Dominance of U.S. Agriculture*. Boulder, CO: Westview Press.

Ward, B. 1989. "Biology: The next frontier." *Sky* (2):85.

Wheeler, D. L. 1988. "Harvard University receives first U.S. patent issued on animals." *Chronicle of Higher Education* 34(32):1.

Winner, L. 1986. *The Whale and the Reactor*. Chicago: University of Chicago Press.

Wysocki, B., Jr. 1987. "Japanese now target another field the U.S. leads: Biotechnology." *Wall Street Journal* 17(12)1,18.

2

Ethical Issues and BST

Paul B. Thompson

The controversy over BST involves disputes about many technical issues: Does milk from cattle treated with BST differ from milk now being produced on dairy farms across the nation? How quickly will dairy farmers adopt BST, and how will it effect economies of scale in milk production? Will milk production shift from the traditional dairy states to new locations? How will rural communities be affected? Are we sure that milk produced by BST-treated cows will be properly metabolized by human consumers?

Answers to these and other technical questions are important because they bear upon questions of responsibility, social justice, and human (and animal) well-being. People assume that the answers to these questions entail that BST should or should not be used, but a complete prescription requires additional premises about responsibility, social justice, and human/animal well-being. There are some applications of these concepts on which our society enjoys a firm consensus, but other applications are notoriously contentious. This chapter will examine some of the issues where controversy over ethical ideals and beliefs may lie just beneath the surface of controversy over facts.

Ethics and Unwanted Consequences of Technical Change

One fact of postmodern society is that decisions by a few individuals to develop and disseminate new technologies can have enormous impact upon society as a whole. Although there are many instances where these impacts are predominantly beneficial, there are few (if any) occasions on which they are universally so. Decisions made far from the rural heartland, in corporate offices or in research facilities, can effectively determine that some producers will have to leave farming, that consumers will be buying new food products, and that rural residents, wildlife, and, indeed, society as a whole (including future generations) will have to cope with pollution and resource depletion. The fact that new technologies produce new benefits, but may also cause unwanted consequences, is the basis for an ethical imperative for decisionmakers to accept responsibility for weighing benefits against unintended consequences (Jonas 1984).

To some extent, market forces discipline these decisions in an open economy. Consumers may reject products that they think unsafe or otherwise undesirable. The discipline of market forces is limited, however. Sometimes the unwanted effects are not borne by the people with the market power to prevent the technology from being adopted. Competitive pressures sometimes lead producers to make choices that ultimately leave everyone worse off. Lack of information sometimes leads people to make poor choices. Though market forces limit the unwanted consequences of new technology, there may still be important issues of fairness and wisdom in accepting the unwanted consequences allowed by these market forces.

The BST case raises questions about three kinds of unwanted consequences. Papers by Comstock (1988), Burton and McBride (1989), Molnar, Cummins, and Nowak (1990), and by Burkhardt (1991) have summarized ethical issues associated with BST. The first group of impacts are felt by dairy farmers who may be forced to adopt BST (or to cease dairy farming) because of competitive pressures (Browne 1987, Buttel 1986). The second includes consequences for nonhuman animals (Comstock 1988). The third includes environmental impacts that may bear upon large numbers of people, extending into future generations (Lanyon and Beegle 1989). Food-safety issues have less to do with unwanted impact than with uncertainty, discussed below. Other chapters in this book detail the potential for unwanted consequences associated with the introduction of BST, and discuss the evidence for concluding whether or not these consequences will occur. Although the range of consequences is quite broad, it is important to see how these three main categories raise different kinds of ethical issues.

1. *Impact upon Dairy Farmers.* There are studies which indicate that BST will increase productivity in herds of those dairy

farmers who adopt it. This increase could make it uneconomical for some other dairy farmers to continue in dairying. Although the specific nature of impacts upon dairy farmers and rural communities varies, the broad ethical issue is one of fairness with regard to those farmers who are adversely affected. Will the costs of technical change be fairly distributed?

2. *Impact upon Dairy Cows.* It is the animals themselves that are most directly affected by BST use. If it is determined that BST harms the dairy cow to which it is administered, this fact will raise the question of whether (and if so, to what extent) the impacts upon animal welfare should militate against use of BST. Do duties to nonhumans impose constraints on technology?
3. *Environmental Impacts of Intensification in the Dairy Industry.* If the introduction of BST helps bring about larger dairy farms, as some studies indicate, BST will be implicated in environmental problems (totally apart from the issue of BST) that dairy farmers are facing. If dairy farms that produce much of their own feed and recycle animal wastes are more ecologically sound than dairy farms which must have feed hauled in and waste hauled out, there may be environmental reasons to promote small-scale farms of the sort that may be threatened by BST. Will the adoption of BST exacerbate existing environmental problems?

Each kind of unwanted consequence is ethically controversial. In most sectors of the economy, producers would not expect to be shielded from the economic consequences of technical change. Farmers are raising a concern more typically voiced by organized labor, as when plant closings or new production lines lead to layoffs. Makers of rollerskates or stainless steel run the risks of being displaced when new technologies appear in their industries; perhaps farmers should, too. Extension of ethical concern to farm animals and to environmental impacts is also hotly debated. While Americans share a strong consensus against cruelty to animals, the stresses imposed upon farm animals in food production have not historically been thought inhumane. Given the assumption that animal husbandry practices carried out for the purpose of producing food for humans are generally acceptable, opponents of BST must show why this particular technology is inhumane, or, alternatively, must show why traditional standards for animal care should be revised. Environmental quality at one time was thought to conflict with important personal rights of autonomy and control over private property. While there is an emerging consensus that these individual rights must sometimes be overridden by environmental concerns, how public policy is to accomplish this goal remains a matter of considerable debate.

Responsibility and Unwanted Impact

Since there is controversy on each of the three points where BST has been linked to unwanted outcomes, it will be useful to look at two ways of framing the ethical issues of responsibility. In some respects, the fundamental issue at stake in the dispute over BST revolves around which of the two philosophical interpretations of responsibility will best guide our collective understanding of the ethical issues involved. Two points are to be observed in comparing these two approaches.

1. The possibility of radically different philosophical approaches to framing the question of responsibility creates many opportunities for misunderstanding among people and groups who view the issue from alternative vantage points.
2. Each vision of responsibility makes its claims about how a democratic society should go about answering questions raised by unwanted consequences of technical change. As such, this dispute cuts to the core of our political culture and begins to involve issues of far greater lasting significance than BST.

The Intentional-action Model

Each of three types of unwanted consequences noted above involves impacts upon individuals or groups who are powerless to avoid being affected. This is clearly the case with respect to farm animals and unborn generations of human beings, and it is true to a more qualified extent for small-scale dairy farmers, too. Each of us must bear the consequences of events we are powerless to control, as when fire and flood destroy our homes, or when faceless economic forces deplete our savings through inflation.

The creation and implementation of BST is not the result of natural causes or of the faceless machinations of the invisible hand, however. BST is on the scene today because a few hundred individuals made research and development decisions over a half decade. The decisions and the actions that followed were undertaken *intentionally*. No one intended that unwanted consequences would occur; some who participated in research did not even care whether BST ever came on the market. Nonetheless, the unwanted outcomes are reasonably predictable consequences of actions that were undertaken on purpose, and this fact marks a crucial distinction between unwanted consequences associated with BST and those that result from natural or faceless social causes.

The individuals and groups that carried out research and development of BST are capable of actions that impose unwanted consequences on others. The question is whether possession of this capacity gives them an unfair or unjust form of power over others. There are two reasons to think that it might.

1. If those who research and develop technologies like BST have unequal advantages over those who bear the unwanted consequences, there is a *prima facie* reason for raising questions about justice.
2. If the researchers who developed BST are agents who have a charge to protect the interests of those who experience unwanted consequences, there may be a betrayal of trust that raises questions of justice.

Examining BST, we find that the companies developing BST have far more economic power than do small dairy farmers. Researchers developing BST in universities do not face the same threats to livelihood faced by small dairy farmers. Animals and future generations are placed into unequal positions relative to present-generation decisionmakers by simple facts of biology. There seems to be *prima facie* reason for raising questions about justice.

These *prima facie* reasons may be reinforced by the fact that many of the scientists who have participated in the development of BST can be thought of as agents for the general public, at least, and perhaps for the farm community, in particular. A great deal of scientific activity in Western democracies is protected against market forces. Few scientists must turn a profit on their labs. Working at public or nonprofit institutions, many scientists aggressively claim a mandate to do science in the public interest whenever research budgets and academic freedom are at issue. If science is to claim this mandate, it must live up to it. Land-grant universities, where much of the BST work has been done, have historically accepted a further mandate to do science that will strengthen the development of rural communities. As such, dairy farmers may have a special claim upon these institutions. Although no one has argued that scientists have a special responsibility to look out for animals, it is not uncommon or unreasonable to think that the scientific community is especially well-placed to look out for the general public interest in environmental quality. In addition to the scientific community's self-appointed role of public service, then, there are additional reasons to think that two of three areas of unwanted consequences fall particularly in the domain of scientific responsibility.

The fact that BST emerges as a technology for which these considerations are relevant *does not settle* the issue in favor of BST's critics. At most they establish a burden-of-proof that must be met by anyone who wishes to impose costs upon farm, animal, and environmental interests. Individuals and groups who would benefit from changes may be able to meet this burden-of-proof, either by argument or by compromise. More narrowly, considerations of fairness and justice establish the right of those who speak for farm, animal, and environmental interests to be heard in the political process. Whether they prevail will depend upon how highly values of equality and trust are rated against needs for food production and economic growth. One might interpret the political debate that has raged over BST as working out an exchange of views in the democratic political process. Whether the handling of BST has been an adequate or effective way for these voices to be heard is a matter that will be taken up below.

The Consequence-valuation Model

The idea that any technical change produces winners and losers invites us to think of a new technology as a social bargain in which there are both costs and benefits. Some technologies, as reflected by their capacity to produce an attractive package of benefits at an acceptable social cost, will be a better bargain than others. The key to evaluating this social bargain lies in identifying and measuring the full range of costs and benefits.

A series of value judgements must be made in order to do this. One must establish the value of consequences, such as an illness or a loss of life, that are only indirectly exchanged for other values (such as consumption or economic growth) in the normal course of events. One must decide how to evaluate consequences that occur at some point in the distant future. One must have some way of being reasonably sure that the accounting is complete, and that important costs (or benefits) have not been omitted. While one should not underestimate the difficulty of making these judgments, the idea that a technology's costs and benefits can be compared with the costs and benefits without the technology provides an attractive way of discharging the imperative of responsibility for technical change.

When applied to BST, the consequence-evaluation model regards adverse, unwanted outcomes as costs weighed against the projected benefits derived from lower milk costs. It is possible that there could be environmental benefits, as well as environmental costs, thus the framework for such an evaluation cannot be set in advance of empirical investigation. While a reduction in milk costs might be slight for an individual consumer, they may be quite large when multiplied by the large number of people who use milk. As such, it is quite possible that the benefits of BST may outweigh the costs. The actual assessment of costs and benefits requires a significant amount of technical expertise. When one adopts a consequence-evaluation model for assessing new technologies, the question of whether BST is an ethically acceptable technology hangs upon the answers to these technical questions.

Notice that when one compares total outcomes from two or more options (at a minimum, the options include BST and no BST), there is no obvious reason why intentional action should enter the picture at all. There are costs and benefits associated with the *status quo*, even when they are the result of faceless economic forces that have evolved through time. If the tradeoffs between cost and benefit for the *status quo* are clearly less attractive than the tradeoffs from technical change, the consequence-evaluation model suggests no reason to place additional significance upon consequences of intentional actions. If an analyst is convinced that equality or autonomy *is* important, these factors enter the consequence-evaluation model as additional sources of value that must be measured if the accounting is to be complete. Assessing the consequence value of a loss of autonomy may prove quite difficult, and it is, in any case, a very different view of how autonomy figures in the imperative of responsibility than that developed in the intentional-action model. The possibility of taking very different approaches to the problem of unwanted outcomes can itself feed policy controversy. It is far easier for two people who have different interpretations of responsibility to talk past each other than it is for them to communicate. Even more fundamentally, some who reject the consequence-evaluation model see it as contrary to traditional values, or as a technocratic response to political problems that require open-ended political dialogue (Sagoff 1988).

These issues are taken up again in the later sections of this chapter. First it is important to survey the ethical importance of consumer concerns that have dominated much of the public discussion of BST.

Ethics and Uncertainty

By its very nature, technical change involves unprecedented events. Although it is possible to predict certain kinds of consequences with reasonable confidence, these predictions are themselves seldom uncontested. The reality of disagreement among alleged experts creates a situation in which a member of the lay public, lacking even the evidence to make informed judgements about who to believe, quite reasonably comes to regard all claims about the likely consequences of technical change with justifiable skepticism. Faced with a line-up of Ph.D.s expressing contradictory claims, the educated layman has no alternative but to bring highly subjective factors to bear in deciding who to believe. Since it is logically impossible for all the experts who espouse contradictory views to be speaking the whole truth, it is very reasonable to question the validity of claims that any expert makes about the true risk of a technology. This skepticism is amplified when alleged experts betray their insensitivity to a layman's dilemma by making statements about the ignorance, irrationality, and emotionalism of public concern, or by describing these concerns as perceived rather than real. It is quite rational to doubt the judgement of a person who seeks your support by calling you a fool.

The unfortunate upshot is that political decisions about technology often become dominated by uncertainty. Technical uncertainty creates an opportunity for experts to disagree. When experts disagree, nonexperts are faced with uncertainty about who to believe. When experts become advocates for a technology, nonexperts have reason to suspect that their advocacy is based upon personal interests. Scientists who criticize public judgement may be attempting to establish positions of power from which they can advance their research interests, their prestige, and ultimately their financial gain. In such a political environment a lay-person may reasonably conclude that information about who has the most to lose or gain by adoption or rejection of technical change is highly relevant to decisions about who to believe (Thompson 1986).

The ethical character of the BST debate changed drastically when claims about the safety of drinking milk from cows treated with BST became a contested issue. Until then, the general issue was how to resolve questions of responsibility for the unwanted consequences of introducing BST. These consequences were of political importance to those who spoke for small farms and animal and environmental interests, but might have been overlooked by many who take an active interest in public affairs. With the advent of controversy over food safety, the potential spectrum of affected parties increased dramatically.

What is even more important is the way that the ethical issue shifted from being one of dealing with unwanted consequences to one of uncertainty. This is a subtle point, but one that should not be missed by students of the BST case. There has been no serious scientific evidence to suggest unwanted health consequences for consumers of BST milk. Consumer groups reacting to the food-safety issue were not reacting to a health risk *per se*, even a negligible one, as in the case of Alar on apples. Consumer groups were reacting to uncertainty, to a problem in deciding who to believe about BST and milk. While the fact that an overwhelming majority of credentialed scientists see no health risk associated with BST milk should (and, in the long-run, probably will) count heavily in favor of BST, the fact that many of these scientists are linked with private corporations and research institutions that stand to gain from sales of BST weighs in against it.

Another subtle distinction should not be missed: the layman does not evaluate the risk of BST as such. The layman must evaluate the risk of choosing the wrong expert. Arrogant behavior and an inability to use common English are relevant pieces of evidence for the decision, "Who should I believe?" In the BST case, consumers may rationally see little cost in being deceived by BST's opponents, and may be comparing this risk to a low-probability/high-consequence risk of being deceived by the majority of voices speaking for the Food and Drug Administration (FDA) and the scientific community.

In understanding the way that ethics bear upon risk and uncertainty, it is crucial to see that all of the consumer's information about the safety of BST is subject to a conditional probability that the source of that information is either ignorant or, worse, dishonest. Since a component of the laymen's risk estimate derives from uncertainty over sources of information, the risk will not be calculated directly from scientific studies. Many authors have taken to describing the difference between risks calculated on the basis of scientific evidence and risks calculated on the basis of corrigibility of human beings who report scientific findings as a distinction between "real" and "perceived" risk (Rowe 1977, Fischhoff et al. 1981, Rescher 1983, Johnson and Covello 1987, Lewis 1990, Glickman and Gough 1990). This choice of words is sometimes unfortunate, for it can be taken to imply that the lay person is responding to extraneous and irrelevant evidence. While it is almost certainly true that many of us do make erroneous and even irrational risk judgments, evidence bearing upon an information source's willingness and ability to report the truth is hardly extraneous or irrelevant to a layman's estimate of food-safety risk.

Given the background of the uncertainty problem faced by food consumers and consumer advocates, it is not surprising that the issue evolved into a debate about the risks of BST and milk. The scientific community has come to view risk issues as an expected-value problem, and this is the way it has approached the food-safety issue for BST. While there are clearly many cases in which the assessment of expected values is the right approach to take for food safety, alternative burden-of-proof approaches may have been a better choice for BST.

Alternative Ethical Approaches to Risk

As can be seen from the discussion of uncertainty, risk problems are extremely diverse and complicated. Two general approaches to risk issues can be outlined, though there are important variations and varieties of each.

1. *Expected-value approaches*

   These approaches assume that risk is defined by at least two key variables: the probability and value of events. If these variables are determined, it is possible to interpret risk as an expected value, i.e., as a set of outcomes anticipated with the given degree of probability.

2. *Burden-of-proof approaches*
   Burden-of-proof approaches associate risk with actions that will ultimately be sanctioned, prohibited, or allowed with qualifications or modifications. Sanction depends upon meeting burdens-of-proof that may or may not require one to assess expected values. If it is the individuals who voluntarily consent to participate in an activity that bear the risks of the activity, it may be possible to permit an action with little or no assessment of expected values.

The longstanding practices of the Food and Drug Administration (FDA) have established a tradition of taking expected-value approaches to the problem of food-borne risks. A vast battery of toxicological tests can be brought to bear in assessing probable harm from using a food substance or additive. The expected-value approach may have seemed especially applicable to BST, for researchers had reason to think that measured risks for BST milk would be identical to those for non-BST milk. This judgement would have been correct if the issue had been one of unwanted consequences, or had risk assessment been performed in a political environment of attempting to find evidence for unwanted consequences. Having found none, one might have presumed that the food-safety issue was at an end.

Within the context of an uncertainty problem, however, expected-value approaches to risk can backfire, providing the lay person with more reason to doubt expert opinion. The reason is simple. From the lay person's perspective, risk assessors are just one more group of self-professed experts. The criterion of assessing the effects of a particular decision on their interests should be applied. If FDA and university scientists are judged to have close working relationships with the companies and scientists who have promoted BST, publication of assuring risk estimates might just as well be interpreted as a ploy, rather than as unbiased reporting of evidence of product safety. Responding to uncertainty problems with technical risk assessments is, to a person unschooled in probability and consequence-evaluation, little more than saying, "Trust me."

Uncertainty issues are politically fractious and intense. It is far from clear that burden-of-proof approaches would have fared better.

It is possible, however, that an agreement to label BST milk might have been interpreted as a gesture of good faith. Those who chose to accept the dominant judgment of the scientific community could capture price savings on milk. Those who chose to reject that judgment for whatever reason are free to do so. The label alters the burden-of-proof that must be offered for BST milk from one of having to trust potentially self-interested parties to one of being in a position of choice.

Companies involved in the manufacture of BST have wished to avoid labels because they fear that labels imply a stigma that will hurt sales. FDA officials have resisted the strategy because they fear that such labels might imply a health benefit where none is known. As such, labels are hardly a panacea. While labels satisfy a burden-of-proof for acceptable risk, requiring labels may have policy implications that are themselves unacceptable.

The emergence of consumer uncertainty over the safety of BST has blown the political issue out of proportion to its significance as a matter of unwanted impacts. While the impacts of BST are important issues, especially if environmental concerns prove to be warranted, they do not pose a threat to democratic institutions for public decision-making in the way that uncertainty issues do. With respect to our long-term prospects for governance of science and technology, BST is probably a minor episode, but we should not overlook the opportunity to learn from it. Some of these larger philosophical issues are taken up in the next, and final, section.

Ethics and Political Consensus

Democratic political theory has evolved around the concept of a social contract. Government arises because free and informed individuals have agreed to be bound by a sovereign political authority. Such agreement is obviously an idealization, perhaps even a metaphor, and the social contract idea has had its share of detractors throughout history. Nevertheless, contract ideals have based the authority of government and public institutions squarely upon consent of the governed. They have consistently opposed metaphysical claims to authority, regardless of source--divine right, textual hermeneutics, or historical materialism. The norm of contractual consent is based upon a recognition that people must find a way to live together, and that stable social expectations about how power will be exercised and how disputes will be resolved are in everyone's interest.

There was a time when the philosophy of scientific enquiry was deeply allied with philosophical arguments for democratic political structures. The ideal of reproducibility was thought to make science an inherently democratic social institution. Under the leadership of Robert Boyle, the British Royal Society sought to make public demonstrability of experimental results the *sine qua non* of scientific inquiry. In Boyle's view, such a standard assured that scientific truth would rest upon propositions whose validity was apparent to everyone, and would create a situation in which public acceptance of validity did not depend upon the scientist's position in political, religious, or academic hierarchies (Shapin and Schaffer 1985). Although he fought bitter battles with Thomas Hobbes, the originator of social-contract theory, Boyle's belief that scientific truth was essentially founded upon public agreement about the incontestability of facts has been imitated by political theorists from Locke to Dewey and Rawls. These social-contract theorists argue that democratic government must emulate scientists' practice of arriving at truth by public debate within a community of inquiry.

BST has tested the social contract. Researchers and private companies have undertaken research and development on BST with the expectation that, if the product finds market acceptance, their efforts will be rewarded. While it is reasonable that they should have expected to deal with some of the unwanted consequences of BST, it was not reasonable to expect that food-safety issues would be among them. The emergence of uncertainty and, in turn, the food-safety issue is evidence of trouble in the contract. It is evidence of a lack of confidence in science and in its institutions. This is a development that should be viewed as quite serious, not only for science, but for the foundations of democratic institutions.

The problem is that both commerce and political decision-making require a certain amount of trust. Trust must, of course, be won, and once won must be preserved. The matter of why the American public is now so skeptical of science institutions and the biotechnology industry is the subject of a much longer discussion than can be attempted here. Whatever the causes, and however just or unjust the suspicion of science might be, the largest and most serious ethical issue associated with BST is the matter of trust. All the other ethical questions feed into this one. The matter of unwanted consequences becomes an ethical issue because we have learned how decisions to research and develop technology can lead to costs that are not born by the people who make the decision or who reap the benefits. It is impossible to avoid all unwanted consequences; stifling technical change has unwanted consequences, too. The ethical problem is that we must trust people who do not bear the costs to look out for the interests of those who do. The issue of uncertainty is even more transparently related to the issue of trust. Even well-educated citizens are placed in a position of needing to trust the statements of scientists on technical issues of risk and safety, but the multiplicity of voices speaking on an issue makes this trust hard to come by. Repressing dissent is hardly the solution, for such a strategy not only violates the basic political liberty of free speech, but is likely to increase public suspicion of scientists' motives.

The idea of a social contract is used to help us understand the range of authority that should be vested in government. The social contract can be seen as a bargain struck among all members of society in which the unrestricted freedom that would exist in a lawless state of nature is given up in exchange for the state's protection of more basic rights, such as life, liberty, and property. It is only recently that scientific research has begun to play an important role in this contract.

There are two reasons for the change. First, science is now seen as essential to government's ability to protect the life and health of its people. Science is instrumental in identifying threats to life in the form of disease or trauma risks that would otherwise appear to be "acts of God." Second, science's role in developing technologies having unwanted consequences means that science can itself be seen as a threat to a person's livelihood, and perhaps even to life. Science is thus put in the position of being both a threat and a guarantor against threats. To some, this may appear to be a case of asking the fox to guard the henhouse.

One way of solving this problem is to build a high wall between that component of science which is in a position of public trust, and that portion of science which is involved in the development of technologies that may produce unwanted consequences. Public science, conducted at nonprofit institutions, would enjoy public confidence. Private science, conducted in the private sector, would be held to the same degree of accountability normally expected of any commercial activity. The flaws in this solution are complicated and subtle. Although they cannot be examined in detail here, it is worth noting four features of contemporary science that make this ideal very difficult to achieve.

1. *Scientific research does not respect the public/private divide.* Often the same basic science underlies regulatory science and technology development alike. Scientists working on key theoretical topics are likely to be colleagues, without regard to their public or private employers, and scientific knowledge is likely to flow back and forth across this divide.
2. *Enforcing a strong separation between public and private science is impractical.* Individual scientists are likely to go back and forth between public and private research, particularly when technologies are closely linked to important theoretical developments, as was the case with BST. Private scientists will have been educated in nonprofit institutions and will naturally maintain friendships there. Any attempt to control such interaction would be an unacceptable intrusion upon the private life of scientists, and enforcement would be enormously expensive.
3. *Public science institutions are finding it necessary to cultivate private sources of research funding.* The costs of scientific research have escalated beyond the capacity of public and foundation resources. Public researchers have adopted strategies such as contract research, joint public/private positions, and the seeking of patent protection for their discoveries as a way to assure that the dollars needed for future research will be secure. Without a massive increase in government expenditure for research, public researchers will continue to depend upon private money for the foreseeable future.
4. *A strong separation between public and private science sectors might well weaken public science.* The three preceeding problems create a situation in which good scientists might well flee public institutions if they were forced to sever contacts with the private sector. For the time being, a post as university professor on a United States campus remains attractive to many of the world's best scientists.

The dilemma, therefore, is deep. The tension between the regulatory- and the technology-stimulating roles of science erodes public trust in science institutions. At the same time, any solution to this problem must be sensitive to the delicate network of personal relationships that make science possible.

Conclusion

The ethical controversy over BST arose because, like many technologies, it may produce some effects that are unwanted. There is no reason to think that the unwanted consequences of BST are particularly dramatic or extreme, but the fact that decisionmakers within public research organizations and private companies can affect others makes these unwanted outcomes an issue of some significance. The significance has escalated, however, because of the food-safety questions that have been raised, and because of the climate of uncertainty that they generated. It is the uncertainty issue that truly threatens to keep BST off the market at this writing, and it is one that the developers of the technology had no reason to expect.

This, in turn leads to the questions of trust that are crucial to democratic institutions. This is not to say that the success or failure of U.S. constitutional democracy hangs upon the BST decision, but this policy problem can be expected to recur in the future with respect to other technology. American society must resolve whether we can expect to develop biotechnology products in an orderly and efficient manner.

Acknowledgments

I would like to thank Professors Jeffrey Burkhardt, H. O. Kunkel, Floyd Byers and Dale Bauman for their comments on early drafts of this chapter and/or presentations on which this chapter is based. Kelly Hancock and David Kriewaldt provided research assistance for which I am also grateful.

References

Browne, W. P. 1987. "Bovine growth hormone and the politics of uncertainty: fear and loathing in a transitional agriculture." *Agriculture and Human Values* IV(1):75-80.

Burkhardt, J. 1991. "Ethics and technical change: The case of BST." Center for Biotechnology Policy and Ethics Discussion Paper CBPE 91-2. College Station, TX: Texas A&M University.

Burton, J. L., and B. W. McBride. 1989. "Recombinant, bovine samatotropin (rbst): Is there a limit for biotechnology in applied animal agriculture?" *Journal of Agricultural Ethics* 2(2):129-160.

Busch, L., W. Lacy, J. Burkhardt, and L. Lacy. 1991. *Plants, Power and Profit*. London: Basil Blackwell.

Buttel, F. H. 1986. "Agricultural research and farm structural change: Bovine growth hormone and beyond." *Agriculture and Human Values* 3:88-98.

Comstock, G. 1988. "The case against bGH." *Agriculture and Human Values* 5(3):36-52.

Doyle, M. P., and E. H. Marth. 1991. "Food safety issues in biotechnology." *Agriculture Biotechnology: Issues and Choices*. West Lafayette, IN: Purdue Univeristy Agricultural Experiment Station. pp 55-66.

Fischhoff, B., S. Lichtenstein, P. Slovic, S. L. Derby, and R. L. Keeney. 1981. *Acceptable Risk*. New York, NY: Cambridge University Press.

Gendel, S., D. Kline, M. Warren, and F. Yates (eds). 1990. *Agricultural Bioethics*. Ames, IA: Iowa State University Press.

Glickman, T. S., and M. Gough (eds). 1990. *Readings in Risk*. Washington, D.C: Resources for the Future.

Harlander, S. K., J. N. BeMiller, and L. Steensen. 1991. "Impact of biotechnology on food and nonfood uses of agricultural products." *Agriculture Biotechnology: Issues and Choices*. West Lafayette, IN: Purdue University Agricultural Experiment Station. pp 41-54.

Johnson, B. B., and V. T. Covello (eds). 1987. *The Social and Cultural Construction of Risk: Technology, Risk, and Society*. Norwell, MA: Kluwer Academic Publishers.

Johnson, G. L., and P. B. Thompson. 1991. "Ethics and values associated with agricultural biotechnology." *Agriculture Biotechnology: Issues and Choices*. West Lafayette, IN: Purdue University Agricultural Experiment Station. pp 121-138.

Jonas, H. 1984. *The Imperative of Responsibility: In Search of an Ethics for the Technological Age*. Chicago, IL: The University of Chicago Press.

Lacy, W. B., L. Busch, and L. R. Lacy. 1991. "Public perceptions of agricultural biotechnology." *Agriculture Biotechnology: Issues and Choices*. West Lafayette, IN: Purdue University Agricultural Experiment Station. pp 139-162.

Lanyon, L. E., and D. B. Beegle. 1989. "The role of on-farm nutrient balance assessments in an integrated approach to nutrient management." *Journal of Soil and Water Conservation* 44:164-168.

Lewis, H. W. 1990. *Technological Risk*. New York, NY: W.W. Norton & Company, Inc.

MacLean, D. E. 1990. "Comparing values in environmental policies: Moral issues and moral arguments." *Valuing Health Risks, Costs, and Benefits for Environmental Decision Making: Report of a Conference*. Washington, D.C: National Academy Press. Pp 83-106.

Marois, J. J., J. J. Grieshop, and L. J. Butler. 1991. "Environmental risks and benefits of agricultural biotechnology." *Agriculture Biotechnology: Issues and Choices*.

West Lafayette, IN: Purdue University Agricultural Experiment Station. Pp 67-80.

Molnar, J. J., K. A. Cummins, and P. Nowak. 1990. Bovine somatotropin: Biotechnology product and social issue in the U.S. dairy industry." *Journal of Dairy Science* 73:3084-93.

Railton, P. 1990. "Benefit-costs analysis as a source of information about welfare." *Valuing Health Risks, Costs, and Benefits for Environmental Decision Making: Report of a Conference*. Washington, D.C: National Academy Press. pp 55-82.

Rescher, N. 1983. *Risk: A Philosophical Introduction to the Theory of Risk Evaluation and Management*. Washington, D.C: University Press of America.

Rowe, W. D. 1977. *An Anatomy of Risk*. New York, NY: John Wiley & Sons.

Sagoff, M. 1988. *The Economy of the Earth: Philosophy, Law, and the Environment*. New York, NY: Cambridge University Press.

Shapin, S., and S. Schaffer. 1985. *Leviathan and the Air-Pump: Hobbes, Boyle, and the Experimental Life*. Princeton, NJ: Princeton University Press.

Sundquist, W. B., and J. J. Molnar. 1991. "Emerging biotechnologies: Impact on producers, related businesses, and rural communities." Pp 23-40 in B. R. Baumgardt, and M. A. Martin (eds). *Agriculture Biotechnology: Issues and Choices*. West Lafayette, IN: Purdue University.

Thompson, P. B. 1986. "Uncertainty arguments in environmental issues." *Environmental Ethics* 8(1):59-75.

PART TWO

Bovine Somatotropin and the Animal

3

BST and Dairy Cow Performance

Lawrence D. Muller

Improving efficiency and profitability has long been a goal in dairy farming. To attain greater efficiency, dairy producers have increasingly adopted new technology. More specifically, feed is a major cost in the production of milk, therefore improving productive or feed efficiency (yield of milk and milk components in relation to the nutritional cost of maintenance) offers a method to improve profitability. Bovine somatotropin (BST) is a technology which improves productive efficiency because increased milk yield allows the proportion of nutrients used for maintenance relative to the nutrients used for milk production to be reduced.

The number of research studies on BST is unprecedented for a new technology. It is estimated that as of December 1989, scientists have conducted in excess of 900 BST studies using more than 20,000 dairy cows throughout the world. The purpose of this chapter is to review the effects of BST on dairy cow performance, milk production and composition, feed intake and efficiency, and factors affecting the animal response. Emphasis will be given to studies published in scientific journals (as of July, 1991) and to review articles. Several review papers (Bauman et al. 1989a, Office of Technology Assessment 1991, Breier et al. 1991, Chalupa and Galligan 1989, Chilliard 1989, Crooker and Otterby 1990, McBride et al. 1988, McGuffey and Wilkinson 1991, Peel and Bauman 1987) have been published and will be emphasized in this chapter. Many other studies have appeared as abstracts from presentations at professional scientific meetings and will likely be published in the near future.

Production Responses

BST use clearly increases milk yield and efficiency of production. This is documented in numerous single-lactation studies of the short- and long-term effects of BST on milk yields, and more recently in reports of longer studies covering as many as eight lactations. The production response is influenced by many physiological, environmental, and management factors. Milk yield generally increases the first few days after BST administration, reaching a maximum within the first week (Chilliard 1989) that is maintained over the remainder of the lactation. With BST, milk production reaches a higher peak yield with an increased persistency of the lactation curve.

Dose Response and Method of Administration. The magnitude of milk response is dependent on the dosage of BST and to a lesser extent on the method of administration. BST must be present each day to maintain an increased milk response. BST is not stored in the body and is rapidly cleared from the bloodstream. To achieve an increase in milk yield, BST must be administered daily or in a prolonged (sustained) release formulation. The prolonged-release formulations contain small volumes and are administered by subcutaneous injections at 2- to 4-week intervals.

Most of the early research involved *daily* BST administration. Chilliard (1989) summarized 21 experiments (20 conducted in United States) with 969 dairy cows. The mean duration of the treatments was 32 weeks with BST administration beginning 5 to 13 weeks postpartum. The milk yield response was dose dependent, increasing 2.8 kg/day with a 5 mg/day dose, 3.9 kg/day with a 10 to 15 mg/day dose, 5.2 kg/day with 20 to 27 mg dose, and 5.6 kg/day with a daily 31 to 50 mg dose. Milkfat, protein, and lactose content were unchanged. A summary of eight trials initiated at 28 to 35 days of lactation revealed 3.8, 5.0 and 5.7 kg/day increases in milk yield with daily BST injections of 12.5, 25, and 50 mg of BST, respectively (Galligan and Chalupa 1989).

A summary of research with a 14-day prolonged-release formulation of BST shows a slightly lower response in milk yield compared to daily administration of BST. A summary of 15 studies with 451 cows receiving BST revealed that administration of 500 mg prolonged-release BST/14 days, (corresponding theoretically to 35 mg BST/day) increased milk yield 4.4 kg/day (Chilliard 1989). Holstein cows (40 on each control and BST) administered a 14-day prolonged-release BST for the remainder of the lactation beginning at 60 days postpartum had a 3.1 kg/day increased fat corrected milk (FCM) compared to controls (Bauman et al. 1989b). The milk yield response was similar in primiparous and multiparous cows. The difference in milk yield between BST and control cows tended to increase as lactation progressed.

In the United States, 870 cows in 15 herds were used to monitor responses to injections of a 14-day prolonged-release form of BST (Thomas et al. 1991). Cows were divided into three stages of lactation with equal numbers of primiparous and multiparous cows in each subgroup assigned to control or BST injection (500 mg/14 days for 12 weeks). The increase in milk yield (Table 3.1) with BST was less for cows in early lactation than for cows administered BST starting at 101 to 189 days postpartum. Multiparous cows had a greater increase in milk production than did primiparous cows. Overall, the milk production increase was 4.8 kg/day; the range among herds was 2.9 to 7.6 kg/cow/day.

TABLE 3.1. Average Milk Yield (kg 3.5 percent FCM/day) for Control and Two-week Prolonged-release BST-treated Cows in Fifteen Herds.[a]

Item	*Control* [b]	*BST* [c]	*Increase of BST Over Controls*
	---kg/cow/day---		
Milk yield	27.2	32.0	+4.8
Parity 1[d]	25.9	30.1	+4.2
Parity 2[e]	28.4	33.9	+5.5
Days-in-milk (DIM)[e]			
57 to 100	30.3	33.9	+3.6
101 to 140	26.8	32.7	+5.9
141 to 189	24.4	29.5	+5.1

[a] Adapted from Thomas et al. (1991).
[b] 433 cows.
[c] 437 cows.
[d] Parity 1, first lactation.
[e] Parity 2, second through eighth lactation.

The initial summary of about 100 cows on formulations (640 and 960 mg) of a 28-day prolonged-release formulation reported increased milk production of 3.1 and 4.2 kg/day, respectively (Chilliard 1989). Cows receiving 960 mg BST/28-day period produced 6.0 kg/cow/day more milk than nontreated controls (McGuffey et al. 1990). A recent North American trial with 190 multiparous Holstein cows from five herds evaluated the lactational performance when a 28-day prolonged-release form of BST was administered in early, mid, or late lactation (Table 3.2). Yield of FCM was increased by each dose of BST in all stages and lactation, and the dose-related improvement was similar regardless of stage of lactation at initiation of treatment. Feed intakes were similar, thus feed efficiency was improved. Weight gain and body condition scores at the end of lactation were lower in BST-treated cows. Within a 28-day injection period, cows increased milk production immediately following administration, reached peak production 4 to 8 days later, followed by a decline in milk to near day-1 levels by week 4 (Figure 3.1). Early-lactation cows compared to mid- and late-lactation cows, tended to maintain higher milk production through the period (Figure 3.1).

TABLE 3.2. Lactational Performance and End-of-lactation Body Condition Scores of Cows Receiving 0, 320, 640, and 960 mg of 28-day Prolonged-release BST Injections Beginning at Three Stages of Lactation.[a]

Item	*Stage of Lactation (DIM)*[b]	*BST (mg/injection)*			
		0	*320*	*640*	*960*
3.5 percent FCM, (kg/cow/day)	28 to 45	29.7	33.5	35.3	35.8
	111 to 166	25.7	29.3	30.3	31.2
	166 to 334	23.0	24.9	27.7	27.3
	All cows (28 to 334)	26.5	29.3	30.8	31.5
Dry-matter intake (kg/cow/day)	All cows	21.5	21.1	21.8	22.2
Body-weight gain (kg/cow/day)	All cows	.57	.39	.40	.35
Ending body-condition score (1-5)	All cows	3.18	2.73	2.74	2.56

[a] Adapted from McGuffey et al. (1991).
[b] Days-in-milk.

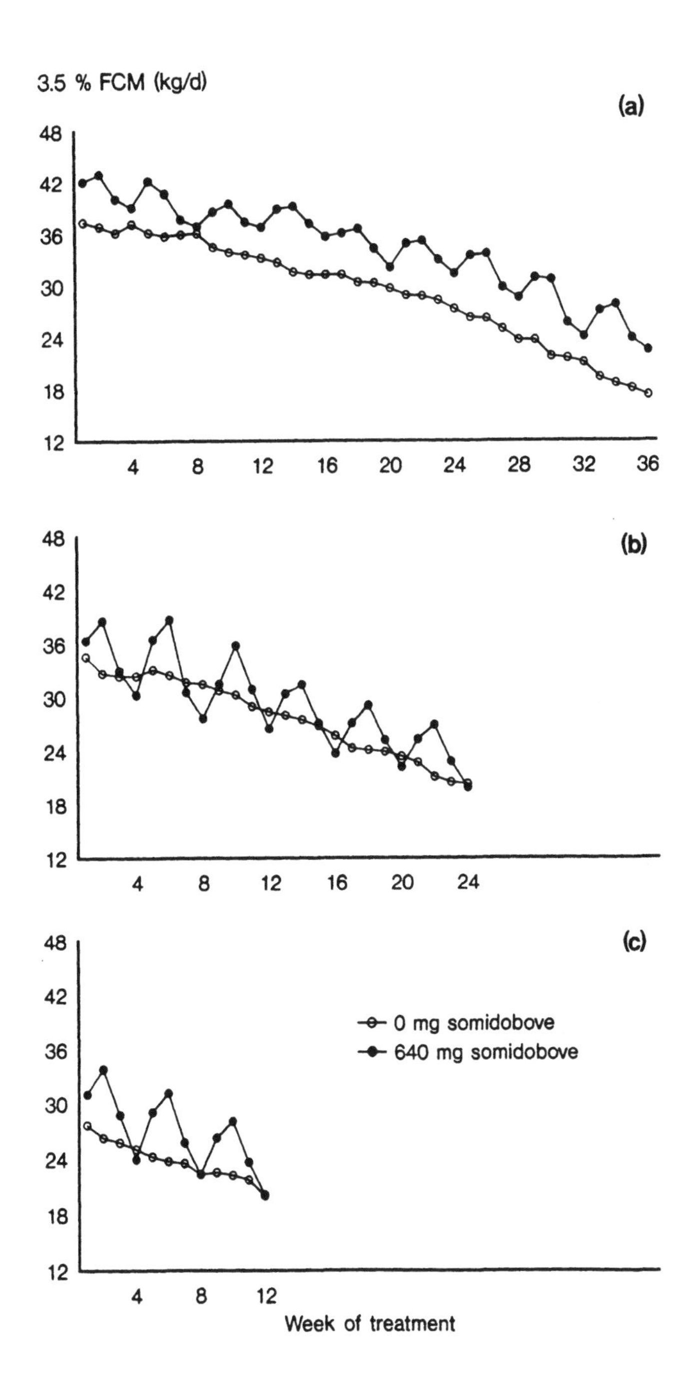

FIGURE 3.1. Average milk production (kg/day) of cows receiving 28-day sustained-release BST calculated for each day of the 28-day administration period using all periods: a) early, b) mid, and c) late. Adapted from McGuffey et al. (1991).

In general, increases in milk yield between the daily, 2-week, and 4- week prolonged-release forms of BST are comparable. During the 14- and 28-day period following administration of the two prolonged-release forms of BST, daily milk yields display more cyclic patterns than when cows are administered BST daily.

Erdman et al. (1991) developed a dose-response summarization model from 44 long-term studies using 3,286 cows. Figure 3.2 shows the 3.5 percent FCM yield of treated and control cows, plotted by dose-response groups from 0-50 mg BST per cow per day. Different methods of administration, including daily injections and prolonged-release forms injected at 14- to 28-day intervals are included, but all are calculated as mg per cow per day. First administration averaged 67 days-in-milk ranging from 28 to 180 days after calving. The average response to BST across experiments was 4.3 kg/day of 3.5 percent FCM. At this time, the dosage to be approved by the Food and Drug Administration (FDA) is unknown, but will likely range between 20 and 40 mg/day and may differ for the daily and prolonged-release forms. Obviously, the payoff from BST use will be dependent on dosage, the magnitude of milk response, and the price of BST.

Results presented are averages for groups of cows. As expected, a range of individual cow response is observed. Variation in cows within BST-supplemented groups is comparable to that of cows in untreated groups (Bauman 1989). One management challenge will be to identify the *low responder* to BST, similar to identifying the *low producer* that is not producing at a profitable level in untreated herds.

Multilactational Response. Most of the research previously cited evaluated the response within a single lactation or a portion of a single lactation. Because of the utilization of body reserves and somewhat lower body condition observed in BST-treated cows during a lactation cycle, knowledge of multilactational responses to BST are important. Several multilactational studies have been conducted but only a few published to date (Annexstad et al. 1990, Huber et al. 1991, McBride et al. 1990, Peel et al. 1989, Phipps et al. 1990). Administration of BST for a second lactation (Annexstad et al. 1990) increased milk yield and feed intake compared to control with the responses similar to those observed during the first lactation of treatment (Soderholm et al. 1988). Increased milk production was maintained when BST was administered during a second lactation (McBride et al. 1990), however, the advantages of BST in terms of improved feed efficiency was reduced

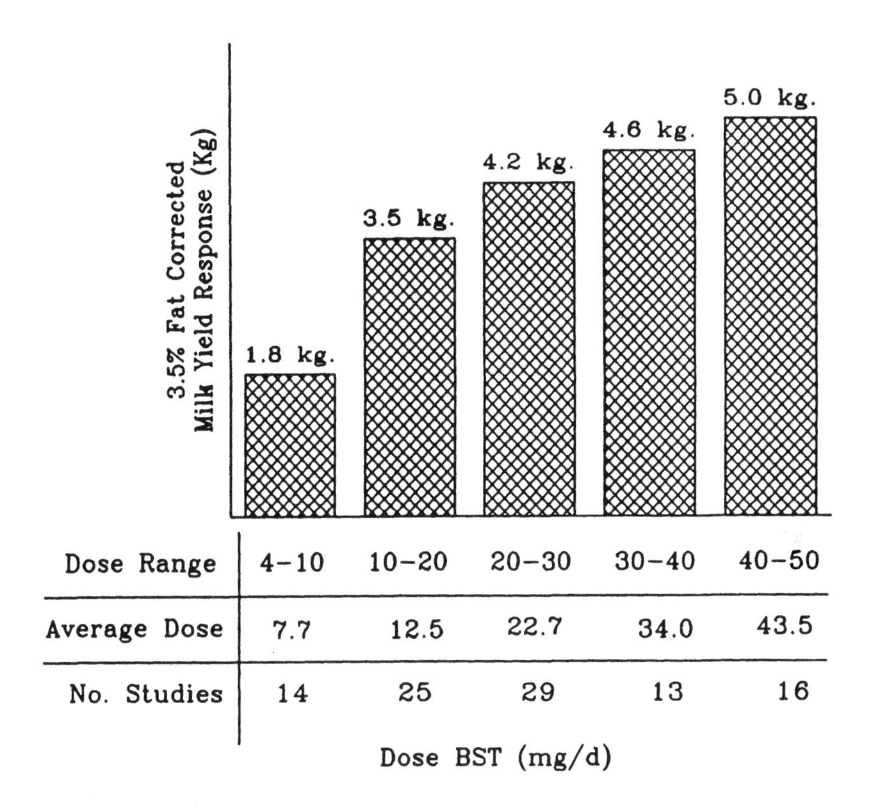

FIGURE 3.2. Milk yield response (3.5 percent FCM/cow/day) of BST-treated cows compared with control cows within the same study. Adapted from Erdman et al. 1991.

during the second lactation. Administration of BST increased milk yield by 4.3 and 3.1 kg/day and increased gross feed efficiency by 8 percent and 6 percent in lactations 1 and 2 (Phipps et al. 1990). Responses to BST for 4 consecutive lactations indicate that milk yields were greater for BST-treated cows during all lactations. Milkfat percent was lower for BST-treated cows during years 2, 3, and 4 (Huber et al. 1991).

The feeding management program may influence the long-term response to BST. When comparing the long-term (three lactations) response to BST under two different feeding programs (Peel et al. 1989), cows fed complete mixed diets and shifted, as individual milk yield decreased, to lower energy-density diets were able to consume sufficient nutrients to maintain a greater milk yield and feed efficiency than did control cows in each of the three lactations. Cows fed concentrate and forage separately with the amount of concentrate based on the milk yield of the previous week, did not reach their maximum milk potential. Milk yield of control and BST-treated cows alike decreased from year to year, and the cows in this study did not respond to BST during the third year (Peel et al. 1989).

Feed Intake. In general, voluntary feed intake increases in BST-treated cows. The increased intake generally begins about 6 to 8 weeks after the start of BST administration, and persists throughout BST use. The magnitude of the increase in feed intake varies depending on the response in milk yield, the method of BST administration, and the energy density of the diet (Chilliard 1989).

In Chilliard's summary, experiments with dairy BST administration reported increased dry-matter intake of 0.6 kg/day and 1.7 kg/day for 10 to 15 mg BST/day and 31 to 50 mg BST/day, respectively. In six studies with the 28-day prolonged-release formulation, dry-matter intake increased from 0.2 to 0.8 kg/day, depending on size of dosage. The study cited in Table 3.2 (McGuffey et al. 1991) reported an 0.7 kg/day increased dry-matter intake with the 960 mg/28 day dosage, whereas milk yield was increased 5.0 kg/day. Administration of lower dosages did not increase feed intake (McGuffey et al. 1991). Cows administered the 500 mg/14-day formulation averaged 1.5 kg/day greater dry-matter intake (Chilliard 1989). Although most studies have reported an increased dry-matter intake with BST, a few studies have reported no change. Lormore et al. (1990) reported no change in feed intake with daily BST administration beginning at 25 days postpartum when milk yield increase was 2.5 kg/cow/day. When all trials are summarized over all BST formulations, the average increase in dry-matter intake is 5 to 8 percent. Thus, cows administered BST generally adjust voluntary intake in a predictable fashion relative to the extra nutrients needed for the increased milk production.

Feed Efficiency and the Lactation Cycle. Based on the previous discussion, milk yield is increased an average of 15 percent with BST, while feed intake is increased 5 to 8 percent. Therefore, average gains in productive or feed efficiency (milk per unit of feed) obtained with BST range from 5 to 15 percent. The gain in production efficiency occurs because the proportion of feed nutrients used by the cow for maintenance is reduced. This is illustrated (Figure 3.3) for a cow which produced 6,818 kg of milk without BST and for a cow which yielded a 20 percent response (8,182 kg milk) to BST (Bauman 1989).

The increased production efficiency is the same for high-producing dairy cows, with or without BST.

The nutrient requirements for maintenance and per unit of milk are not changed by BST. Cows administered BST have a greater total nutrient requirement because of more milk produced (Figure 3.3). They have higher production efficiencies, however, because body maintenance represents a smaller proportion of the total nutrient intake and requirement.

Energy balance and utilization in BST-treated cows, compared to controls, differ during the typical lactation cycle. A typical pattern of milk yield, net energy intake, and net energy balance during a lactation cycle is shown in Figure 3.4. Milk production peaks at 4 to 6 weeks postpartum and decreases through the remainder of the lactation cycle.

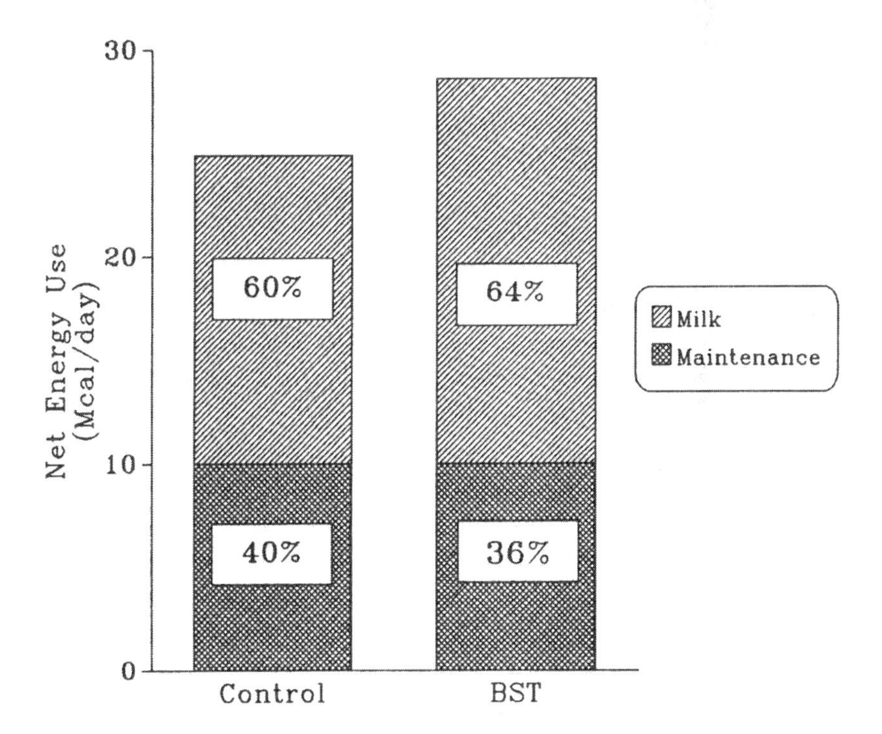

FIGURE 3.3. Efficiency gains by reapportioning the nutrients used for milk production and for body maintenance. Assumes control cow produced 6,818 kg (15,000 lb) of milk in a lactation. The use of BST increases milk yield by 20 percent. From Bauman (1989).

Feed intake (net-energy intake) increases gradually in the first few weeks of lactation and peaks about 10 weeks postpartum or 4 to 6 weeks after peak milk. Therefore, dairy cows are usually in negative energy balance during the first 10 to 12 weeks of the lactation cycle and body reserves are utilized to provide energy for the peak milk

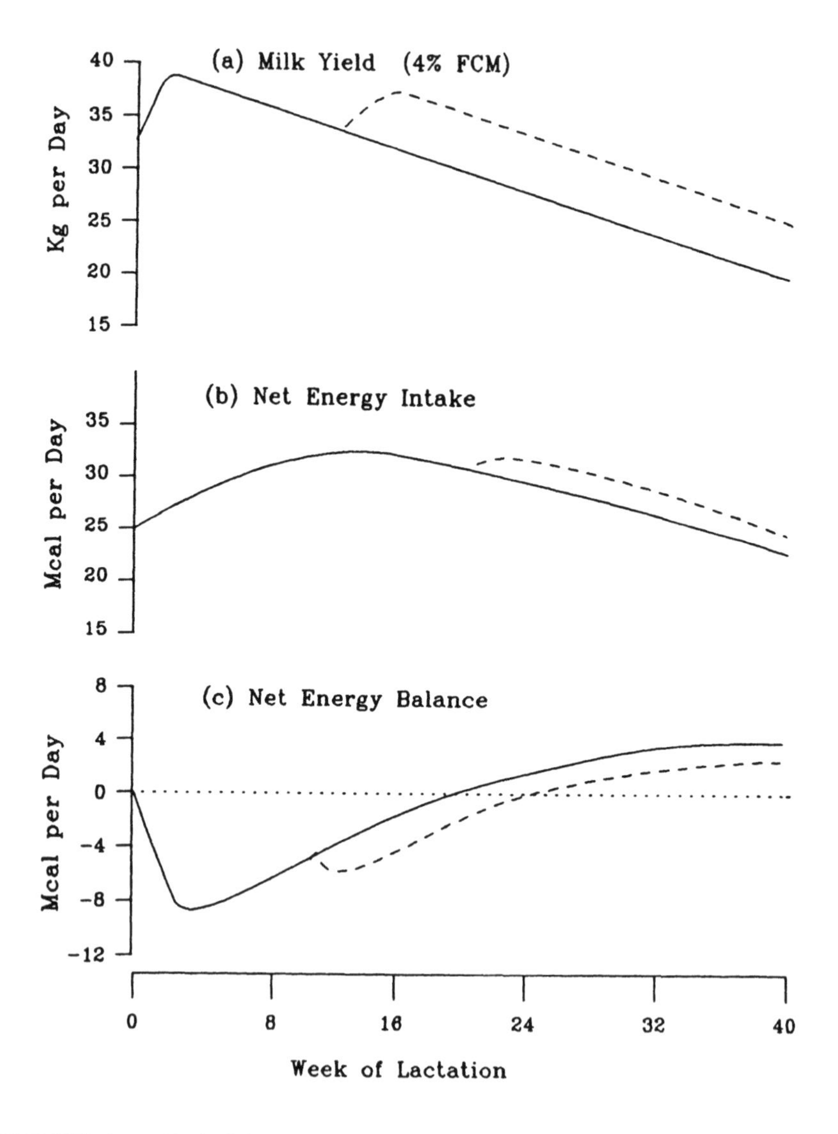

FIGURE 3.4. Typical lactation cycle (solid lines) of high-producing dairy cows, illustrating milk yield (a), net energy intake (b), and net energy balance (c). Dotted lines depict typical response during a lactation cycle if BST administration is initiated at 12 weeks postpartum.

yield. A typical response to BST is plotted with the dotted lines in Figure 3.4. After BST administration, milk yield will increase to a new peak. BST increases the persistency of the lactation curve, thus after peak milk yield, production of treated cows tends to decrease at a slower rate compared to untreated cows. The increased feed intake (energy intake) occurs approximately 6 to 8 weeks after the new lactation peak, and the high-producing dairy cow may enter a brief period of negative energy balance until energy intake increases to meet net energy output (milk). A decreased energy balance does not necessarily mean cows are returning to negative energy balance (as shown in Figure 3.4). They may be less positive, depending on the days postpartum that BST administration is initiated. In general, a reduced body-weight gain and lower body condition score have been reported during a lactation cycle in many BST studies (see, e.g., McGuffey and Wilkinson 1991). The action of BST is to partition nutrients from body repletion to milk production, thus cows administered BST will be expected to gain less weight and body condition than nontreated cows. Replenishment of body-energy stores is needed in late lactation and during the dry period to ensure adequate body reserves for high milk production in the subsequent lactation. Cows that regained body condition during late lactation (Soderholm et al. 1988) responded with 4.8 kg/day more milk production during a second consecutive lactation with 20.6 mg BST/day (Annexstad et al. 1990), a milk response comparable to the previous lactation. Dairymen will need to monitor body condition and management of feeding programs to ensure that cows will respond to BST in subsequent lactations.

Milk Composition. Many studies have examined the effect of BST administration on milk composition. The overall composition of milk (fat, protein, lactose, minerals, cholesterol, and vitamins) is not substantially altered by BST supplementation and is within the range found in untreated cows (see reviews by Barbano and Lynch 1989, Chalupa and Galligan 1989, McBride et al. 1988, Peel and Bauman 1987, Vanden Berg 1991). The concentration of casein, a major milk protein, is not altered. There can be an increased milkfat content during the first few weeks of BST administration. The temporary shift in milkfat content relates to nutritional status; cows in negative energy balance may produce milk having a higher fat content for a brief period due to greater mobilization of body fat. BST may cause an increase in the relative amount of long-chain fatty acids in milk

(Vanden Berg 1991). The meat derived from BST-treated cows has a lower fat content but is otherwise similar to that from untreated cows.

Factors Affecting the Production Response

Stage of Lactation. In general, the milk response to BST is smaller with administration of BST in early lactation than when administration is started after peak milk yield is attained. Data from a study with 870 cows at 15 locations (Table 3.1) show a lower milk response when BST administration was started from 57 to 100 days-in-milk (DIM) compared to starting during the last two-thirds of lactation. In contrast, cows administered a 28-day prolonged-release form of BST beginning 28 to 45 DIM had a milk response similar to administration beginning after day 111 of lactation (Table 3.2). It is likely that the use of BST would typically occur the last three-fourths of a lactation cycle, after peak milk yield, because of the somewhat lower and more variable response to BST treatment in early lactation.

Breeds. Lactational responses to BST have been reported for all dairy breeds that have been studied. West et al. (1990) reported a similar pattern of increased milk yield and feed-intake response to BST treatments in Holsteins and Jerseys. Pell et al. (1988) reported that the response of Jerseys to BST treatment was of a similar magnitude to the response of Holsteins.

Parity (lactation number). The milk yield response by multiparous cows has often been greater than that of primiparous (first lactation) cows (Table 3.3) (Chilliard 1989, Peel et al. 1989, Thomas et al. 1991). However, some studies cited by Chilliard (1989) report no differences. This lower and variable response to BST may be associated with the fact that, in addition to producing milk, heifers are still growing during the first lactation. The response of heifers to BST may be related to their growth prior to calving.

Climatic Stress. High environmental temperatures and relative humidities can lead to reduced performance and altered physiologic status in lactating dairy cows. Environmental tolerance of a lactating cow administered BST has been a concern and has been reviewed (Hartnell and Collier 1989). Cows receiving BST in warm environments generally respond with increased milk yield, but to alesser extent than cows not heat-stressed (Zoa-Mboe et al. 1989). Cows treated with

TABLE 3.3. Effect of Stage of Lactation on Average Milk-yield Response (kg 3.5 percent FCM/cow/day) of Cows Receiving BST.[a]

	Milk Yield		*Difference*	
Lactation	*Control*	*BST*	*Yield*	*(%)*
	------------(*kg 3.5% FCM/cow/day*)------------			
Primiparous				
57-100 DIM[b]	27.4	30.6	3.2	11.7
101-140 DIM	25.6	30.8	5.2	20.3
141-189 DIM	24.8	29.0	4.2	16.9
Multiparous				
57-100 DIM	33.3	37.3	4.0	12.0
101-140 DIM	28.0	34.6	6.6	23.6
141-189 DIM	23.9	29.9	6.0	25.1

[a] Adapted from Thomas et al. (1991).
[b] Days-in-milk at initiation of BST administration.

BST under heat stress apparently dissipate the increased heat load through increased respiratory and skin vaporization. Thus, BST use does not appear to affect heat tolerance. Assuming good feed, water, and environmental management, the evidence suggests that BST can be used under heat-stress conditions (Hartnell and Collier 1989).

Genetic Potential/Milk Production Level. In general, administration of BST to *well-managed* cows will increase milk yield, regardless of the genetic potential (Leitch et al. 1990, Nytes et al. 1990). As expected, response varies among herds and within herds, as is the case with untreated herds (Thomas et al. 1991). There are conflicting reports concerning the influence of phenotype on a cow's ability to respond to BST. Thomas et al. (1991) utilized data from 15 trials to calculate correlations between pretreatment milk yield and change in milk yield during the treatment period for control and BST-treated cows. Similar correlation coefficients were noted, suggesting that milk yield of a cow cannot be used to reliably predict her ability to respond to BST. In contrast, phenotypic potential for production was significant in explaining variation in response to treatment; cows of low and medium production potentials responded more to treatments than did their counterparts with higher production potential (Leitch et al. 1990).

Management and Nutrition. Quality of managment and nutrition may be the major factors affecting the magnitude and duration of milk yield response to BST (see Figure 3.5). A poorly managed dairy farm may be expected to achieve little or no benefit from BST use, because other factors will likely limit the ability of the cow to produce milk. Maximal response, on the other hand, would be expected from a herd with excellent management. In a summary of 45 field studies conducted in the United States, Europe, and Africa, a correlation of 0.58 was observed between pretreatment group milk yield (an indication of quality of management) and the milk yield response to BST (Peel et al. 1989). Deficiencies in environmental, nutritional, and milking-management programs may have attributed to the lack of response to BST (Mollett et al. 1986).

Results of a multisite study discussed by Crooker and Otterby (1990) further demonstrate the variation among dairy herds in response to BST. In six dairy farms following the same BST-treatment regimen, the overall milk response, relative to control cows, ranged from 5.5 to

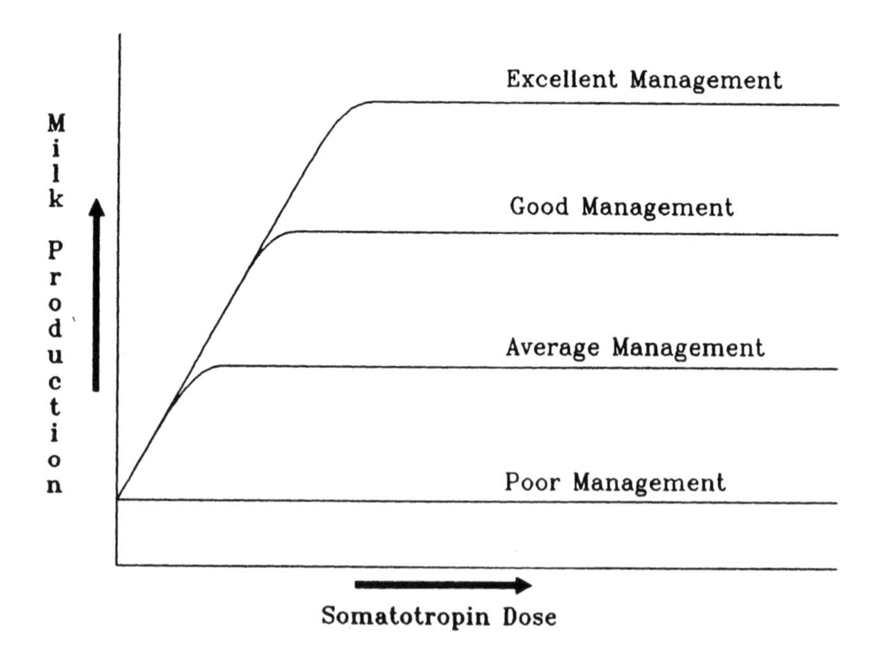

FIGURE 3.5. Effect of quality of management on milk response of dairy cows receiving bovine somatotropin. Adapted from Bauman (1989).

18.3 percent for treated primiparous cows and 5.1 to 29.4 percent for treated multiparous cows. A closer examination of the data suggests that body-condition of the cows prior to the start of treatment had an effect on the milk response. Herds with higher body-condition scores tended to have a higher response to BST. In contrast, Peel et al. (1989) summarized 23 BST studies and found no association between average body condition scores before treatment and the subsequent response to BST. However, it seems logical that achieving appropriate body condition prior to calving and to BST administration is important.

The nutritional-management program is a major factor influencing the response to BST. Nutrient utilization by the dairy cow is not altered by BST administration. The digestibility of dietary components and the energy and nutrients required for maintenance and to synthesize milk (metabolic efficiency) are similar in nontreated controls and BST-treated cows (Tyrrell et al. 1988). Three experiments have examined the partial efficiency of metabolizable energy use for milk yield (Kirchgessner et al. 1991, Sechen et al. 1989, Tyrrell et al. 1988) and all have reported similar partial efficiencies for BST-treated and control cows. Therefore, the current National Research Council (1989) nutrient recommendations for lactating dairy cattle are applicable to the BST-treated cow (Chalupa and Galligan 1989). BST-treated cows consume more feed, therefore, the dairy farmer must utilize management techniques which will increase feed intake.

Research studies which tend to confirm that the 1989 National Research Council dietary recommendations for high-producing cows are applicable to BST-treated cows have been published (Chalupa and Galligan 1989, Lormore et al. 1990, McGuffey et al. 1990). Administration of BST causes a shift in partitioning of available energy toward milk yield at the expense of body tissue, suggesting that an increased energy-dense diet may be needed. Addition of calcium salts of long-chain fatty acids to increase energy density of the diet resulted in an enhanced milk yield in BST-treated cows (Schneider et al. 1990). Other sources of supplemental fat would likely provide similar results. The previously discussed responses to long-term administration of BST (Peel et al. 1989) illustrate that the nutritional-management program and management of body reserves by the dairyman will have a large influence on the lactational response to BST. Clearly, additional research publications and investigations are needed in this area.

Conclusions

Nearly 1,000 world-wide studies involving more than 20,000 dairy cows have confirmed that administration of BST can boost milk yields 10 to 20 percent or about 4 to 5 kg/day, with 5 to 10 percent less feed per unit of milk. An average daily BST dose between 10 and 30 mg/day appears to be optimum. To be effective, BST must be administered daily or by a prolonged-release formulation. Productive efficiency (milk per unit of feed offered) is increased approximately 5 to 15 percent because the proportion of nutrients used for body maintenance is reduced and body-tissue gains tend to decrease. Thus, the use of BST represents an advantage of reduced costs per unit of milk, but overall profitability is dependent on milk price, feed and BST costs, and other economic factors. Milk composition and nutritional quality of milk from BST-treated cows does not differ from that of control cows. A larger response to BST has been generally reported during the last 200 to 230 days of the 305-day lactation cycle and has usually been somewhat greater for multiparous than for primiparous cows. Positive responses have been observed for all breeds studied. BST-treated cows and nontreated cows respond similarly under heat stress. Administration of BST to well managed cows will increase milk yield regardless of genetic potential or milk production levels. Quality of management and nutrition programs will be a major factor affecting the magnitude of milk response to BST.

The research on BST is extensive, but a sizeable amount is yet to be published. When available, this new information may provide additional insight into the management of the BST-treated herd, and may suggest some modest changes to the conclusions above.

Acknowledgments

I am grateful to Dr. Roger Hemken, Department of Animal Sciences, University of Kentucky, and Dr. Al Rakes, Department of Animal Sciences, North Carolina State University, for their helpful review comments.

References

Annexstad, R. J., D. E. Otterby, J. G. Linn, W. P. Hansen, C. G. Soderholm, and J. E. Wheaton. 1990. "Somatotropin treatment for a second consecutive lactation." *Journal of Dairy Science* 73:2423-2436.

Barbano, D. M., and J. M. Lynch. 1989. "Milk from BST-treated cows: composition and manufacturing properties." Pp 9-18 in *Advanced Technologies Facing the Dairy Industry: BST.* Ithaca, NY: Cornell Cooperative Extension Animal Science Mimeograph Series #133.

Bauman, D. E. 1989. "Biology of bovine somatotropin in dairy cattle." Pp 1-8 in *Advanced Technologies Facing the Dairy Industry: BST.* Ithaca, NY: Cornell Cooperative Extension Animal Science Mimeograph Series #133.

Bauman, D. E., F. R. Dunshea, Y. R. Boisclair, M. A. McGuire, D. M. Harris, and K. L. Houseknecht. 1989a. "Regulation of nutrient partitioning: homeostasis, homeorhesis and exogenous somatotropin." Keynote lecture in F. A. Kallfelz (ed). *Seventh International Conference on Production Disease in Farm Animals.* Ithaca, NY: Cornell University.

Bauman, D. E., D. L. Hard, B. A. Crooker, M. S. Partridge, K. Garrick, L. D. Sandles, H. N. Erb, S. E. Franson, G. F. Hartnell, and R. L. Hintz. 1989b. "Long-term evaluation of a prolonged-release formulation of N-methionyl bovine somatotropin in lactating dairy cows." *Journal of Dairy Science* 72:642-651.

Breier, B. H., P. D. Gluckman, S. N. McCutcheon, and S. R. Davis. 1991. "Physiological responses to somatotropin in the ruminant." *Journal of Dairy Science* 74(Suppl 2): 20-34.

Chalupa, W., and D. T. Galligan. 1989. "Nutritional implications of somatotropin for lactating cows." *Journal of Dairy Science* 72:2510-2524.

Chilliard, Y. 1989. "Long-term effects of recombinant bovine somatotropin (rBST) on dairy cow performances: a review." Pp 61-87 in K. Sejrsen, M. Vestergaard, and A. Neimann-Sorensen (eds). *Use of Somatotropin in Livestock Production.* New York, NY: Elsevier Applied Science.

Crooker, B. A., and D. E. Otterby. 1990. "Managing the BST treated cow." Pp 59-83 in *Proceedings of 51st Minnesota Nutrition Conference.* Bloomington, MN: University of Minnesota.

Erdman, R. A., L. W. Douglass, and M. A. Varner. 1991. "How cows respond to BST." *Hoard's Dairyman* 136(9):406.

Galligan, D. T., and W. Chalupa. 1989. "Economic effects of BST on your farm." Pp 49-56 in *Advanced Technologies Facing the Dairy Industry: BST.* Ithaca, NY: Cornell Cooperative Extension Animal Science Mimeograph Series #133.

Hartnell, G. F., and R. J. Collier. 1989. "Will dairy cows response to BST in the tropics and subtropics?" Pp 34-43 in *Proceedings of 26th Florida Nutrition Conference.* Gainesville, FL: University of Florida.

Huber, J. T., J. L. Sullivan, S. Willman, M. Arana, M. Derrigan, C. DeCorte, R. G. Hoffman, and G. F. Hartnell. 1991. "Response of Holstein cows to bi-weekly sometribove injections for 4 consecutive lactation." *Journal of Dairy Science* 74(Suppl 1):211 (Abstract).

Kirchgessner, M., W. Windisch, W. Schwab, and H. L. Muller. 1991. "Energy metabolism of lactating dairy cows treated with prolonged-release somatotropin or energy deficiency." *Journal of Dairy Science* 74(Suppl 2):35-43.

Leitch, H. W., E. B. Burnside, and B. W. McBride. 1990. "Treatment of dairy cows with recombinant bovine somatotropin: Genetic and phenotipic aspects." *Journal of Dairy Science* 73:181-190.

Lormore, M. J., L. D. Muller, D. R. Deaver, and L. C. Griel, Jr. 1990. "Early lactation responses of dairy cows administered bovine somatotropin and fed diets high in energy and protein." *Journal of Dairy Science* 73:3237-3247.

McBride, B. W., J. L. Burton, and J. H. Burton. 1988. "The influence of bovine growth hormone (somatotropin) on animals and their products." *Research and Development in Agriculture* 5:1-21.

McBride, B. W., J. L. Burton, J. P. Gibson, J. H. Burton, and R. G. Eggert. 1990. "Use of recombinant bovine somatotropin for up to two consecutive lactations on dairy production traits." *Journal of Dairy Science* 73:3248-3257.

McGuffey, R. K., R. P. Basson, D. L. Snyder, E. Block, J. H. Harrison, A. H. Rakes, R. S. Emery, and L. D. Muller. 1991. "Effect of somidobove sustained release administration on the lactation performance of dairy cows." *Journal of Dairy Science* 74:1263-1276.

McGuffey, R. K., and J. I. D. Wilkinson. 1991. "Nutritional implications of bovine somatotropin for the lactating dairy cow." *Journal of Dairy Science* 74:(Suppl 2): 63-71.

McGuffey, R. K., H. B. Green, and R. P. Basson. 1990. "Lactation response of dairy cows receiving bovine somatotropin and fed rations varying in crude protein and undegradable intake protein." *Journal of Dairy Science* 73:2437-2443.

McGuffey, R. K., H. B. Green, R. P. Basson, and T. H. Ferguson. 1990. "Lactation response of dairy cows receiving bovine somatotropin via daily injection or in a sustained release vehicle." *Journal of Dairy Science* 73:763-771.

Mollett, T. A., M. DeGeeter, R. L. Belyea, and R. A. Youngquist. 1986. "Biosynthetic or pituitary extracted bovine growth hormone induced galactopoiesis in dairy cows." *Journal of Dairy Science* 69(Supl 1):118 (Abstract).

National Research Council. 1989. *Nutrient requirements of dairy cattle*. 6th revised edition. Washington, D.C: National Academy of Science.

Nytes, A. J., D. K. Combs, G. E. Shook, and R. D. Shaver. 1990. "Response to recombinant bovine somatotropin in dairy cows with different genetic merit for milk production." *Journal of Dairy Science* 73:784-791.

Office of Technological Assessment. 1991. *U. S. Dairy Industry at a Crossroad: Biotechnology and Policy Choices.* Special Report OTA-F-470, pp. 33-50. Washington, D.C: U. S. Government Printing Office.

Peel, C. J., and D. E. Bauman. 1987. "Somatotropin and lactation." *Journal of Dairy Science* 70:474-486.

Peel, C. J., D. L. Hard, K. S. Madsen, and G. de Kerchove. 1989. "Bovine somatotropin: mechanism of action and experimental results from different world areas." Pp 9-18 in *Meeting the Challenge of New Technology*. Monsanto

Technical Symposium preceding the Cornell Nutrition Conference. Animal Sciences Division, Monsanto Agricultural Co., St. Louis, MO.

Pell, A. N., D. S. Tsang, M. T. Huyler, B. A. Howlett, J. Kunkel, and W. A. Samuels. 1988. "Response of Jersey cows to treatment with sometribove, USAN (recombinant methionyl bovine somatotropin) in a prolonged-release system." *Journal of Dairy Science* 71(Suppl):206 (Abstract).

Phipps, R. H., R. F. Weller, N. Craven, and C. J. Peel. 1990. "Use of prolonged-release bovine somatotropin for milk production in British Freisian dairy cows. I. Effect on intake, milk production and feed efficiency in two consecutive lactations of treatment." *Journal of Agricultural Science* 115:94-104.

Schneider, P. L., D. Sklan, D. S. Kronfeld, and W. Chalupa. 1990. "Responses of dairy cows in early lactation to bovine somatotropin and ruminally inert fat." *Journal of Dairy Science* 73:1263.

Sechen, S. J., D. E. Bauman, H. F. Tyrrell, and P. J. Reynolds. 1989. "Effect of somatotropin on kinetics of nonesterified fatty acids and partition of energy, carbon, and nitrogen in dairy cows." *Journal of Dairy Science* 72:59-67.

Soderholm, C. G., D. E. Otterby, J. G. Linn, F. R. Ehle, J. E. Wheaton, W. P. Hansen, and R. J. Annexstad. 1988. "Effects of recombinant bovine somatotropin on milk production, body composition, and physiological parameters." *Journal of Dairy Science* 71:355-365.

Thomas, J. W., R. A. Erdman, D. M. Galton, R. C. Lamb, M. J. Arambel, J. D. Olson, K. S. Madsen, W. A. Samuels, C. J. Peel, and G. A. Green. 1991. "Responses by lactating cows in commercial dairy herds to recombinant bovine somatotropin." *Journal of Dairy Science* 74:945-964.

Tyrrell, H. F., A. C. G. Brown, P. J. Reynolds, G. L. Haaland, D. E. Bauman, C. J. Peel, and W. D. Steinhour. 1988. "Effect of bovine somatotropin on metabolism of lactating dairy cows: Energy and nitrogen utilization as determined by respiration calorimetry." *Journal of Nutrition* 118:1024-1030.

Vanden Berg, G. 1991. "A review of quality and processing suitability of milk from cows treated with bovine somatotropin." *Journal of Dairy Science* 74(Suppl 2): 2-11.

West, J. W., K. Bondari, and J. C. Johnson, Jr. 1990. "Effects of bovine somatotropin on milk yield and composition, body weight, and condition score of Holstein and Jersey cows." *Journal of Dairy Science* 73:1062-1068.

Zoa-Mboe, A., H. H. Head, K. C. Bachman, F. Baccari, Jr., and C. J. Wilcox. 1989. "Effects of bovine somatotropin on yield and composition, dry matter intake, and some physiological functions of holstein cows during heat stress." *Journal of Dairy Science* 72:907-916.

4

Management of BST-Supplemented Cows

R. A. Patton and C. W. Heald

Early in the development of recombinant-BST technology it was recognized that BST use on commercial dairy farms would be enhanced by superior animal feeding and care (Kalter 1984, Mix 1987). Bauman (1987) hypothesized a relationship between management and response to BST as presented in Figure 3.5 (see Chapter 3). This relationship suggests that farm operators whose management practices promote high production will benefit most from use of BST. Peel et al. (1989) present evidence from various locations that support this hypothesis, although they suggest that the biological stimulus of BST is universal in short-term trials. BST has been commercialized in several countries and is widely used under a variety of management conditions in Mexico and Brazil. In the United States, most of the published data on BST results is from clinical trials conducted with university herds under guidelines established by the Food and Drug Administration (FDA) to demonstrate safety and effectiveness of BST at various dosage levels. Clinical trials have been conducted in different geographical locations so that environmental conditions varied widely; however, management practices have been similar in all such trials. Field trials on commercial farms have also been conducted to gain information under a range of management conditions. Such data is now being published. The shortage of specific research addressing management practices that augment BST response necessitates that much of the discussion on this subject be speculative and based on our knowledge of the care and handling of untreated cows. Additional limitations confining our discussion of management and BST are that commercial dosages have not been established and the cost per dose is not yet known. Dosage impacts the magnitude of milk response (Erdman et al. 1990), and its price determines on which cows, and how frequently, BST will be used within a herd (Marsh et al. 1988, Yonkers et al. 1989).

This paper will: (1) discuss management strategies successfully employed with BST, (2) speculate on management factors that may impact BST response, (3) identify areas where management changes may optimize profit with BST, and (4) define when BST might best be used in an overall management strategy designed to maximize dairy farm profitability.

Biologic Action

Central to the discussion of BST-treated cows is the biologic action of somatotropins. Although this action is incompletely understood, from a management perspective the following facts are relevant:

1. BST directs nutrients from other body tissues toward the mammary gland. (For reviews see Peel and Bauman 1987, Hart 1988, McBride et al. 1988). Physiologically, BST increases blood flow through the udder, decreases nonmammary use of blood sugar, decreases body-fat production and increases body-fat breakdown, and coordinates mineral balance between bone and extra-cellular fluids.
2. Two to five weeks after injections begin, BST causes an increase in dry-matter intake (DMI) sufficient to provide extra nutrients for increased milk production (Bauman et al. 1985, Phipps et al. 1990, Hartnell et al. 1991).
3. BST causes a decrease of body fat that continues until a balance of additional dietary energy is provided by BST-induced DMI increase (Tyrrell et al. 1988, Bauman et al. 1989, Hartnell et al. 1991).
4. BST has no effect on the digestive- or partial-efficiency of nutrient utilization (Tyrrell et al. 1988). The BST-treated cow appears similar in this respect to the untreated cow producing at a similar level.

Management strategies that promote the actions of BST, or at least do not impede those actions, should therefore result in the greatest milk production response to BST supplementation.

BST Response on the Individual Farm

To manage BST on commercial dairy farms, it is necessary to measure response to BST in terms of milk yield and profitability. At this point, the latter is difficult, given the record-keeping as practiced by most dairy farmers. Keeping detailed records, then, is the first major management change that many dairymen must make to fully utilize BST technology. To capitalize on the value of BST, applied research must define the circumstances under which BST can be a profitable adjunct to present management systems. There are few investigations providing assistance in identifying the management information that needs to be captured, or suggesting how the needed management information should be captured. Some examples follow. Muller (Chapter 3) suggests that body condition is a consideration in the management of BST-treated cows. Further, it is imperative that body-condition scoring (BCS) be done by qualified individuals and that this information be recorded with other management information. Statewide Dairy Herd Improvement Association (DHIA) programs to capture BCS are just beginning. Maximal feed DMI is critical to maximal milk yield, with or without BST. Few dairymen monitor DMI and fewer still evaluate DMI data over time. A system of more-detailed cost accounting needs to be implemented by dairy farmers, and methods of evaluating the cost of management changes need to be developed. Commercially available accounting systems for farm use need to be flexible enough to permit management analysis. Technical people with the knowledge, capability, and desire to work with dairy managers and develop management strategies for individual farms are needed with or without the use of BST. Leadership in the dairy industry will need to develop ways to address these important needs if technologies such as BST are to be used to greatest benefit.

BST technology will require a different, if not more-sophisticated, set of management records than currently used. People who work with dairy clients--veterinarians, feed dealers, DHIA supervisors, consultants, and Extension agents (as well as the pharmaceutical companies producing BST)--must begin now to develop strategies that will help their clients collect and record the information needed in order to evaluate BST and other technologies as management tools in individual herds. Dairy farmers must be shown a beneficial response from the use of new practices before wide adoption can be expected.

Although the measurement of milk-production response is easy in a research setting, where pairs of cows, one supplemented and one not, are maintained, measurement of milk response in commercial settings is more difficult. Variations among farms in daily milk production, age and size of cow, stage of lactation, breeding, disease, and environment confound comparisons. Typical cow variation is discussed by Peel et al. (1989). Often facilities and management styles do not accommodate control and treated groups. When all eligible cows in a herd are treated, evaluation of BST response is less precise and involves analysis of (1) differences between pre- and post-treatment production over some defined length of time; (2) a change in the shape of the lactation curve relative to some predicted production curve; or (3) a change in the bulk tank reading. Such results can be further confounded by pulsatile milk response that is the result of prolonged-release formulations of BST. That is, daily milk rises rapidly post-treatment and slowly falls to near pre-treatment production levels between BST treatments. Comparison of milk production data is also complicated by large natural variations in milk and milk-component yields. The common practice of random measurement of milk production on a monthly basis is probably not sufficient to define the BST response adequately. Cows found unsuitable for treatment do not serve as reliable controls, particularly because of their dissimilarity in stages of lactation or perhaps a health condition.

Currently, there is no published research to say what types of measurements adequately define milk-production response when no cows are maintained as controls. Although some DHIA groups are beginning to assess the level of BST response that can be predicted from production data, others have not begun to monitor the process. Given our present practices and level of knowledge, it is likely that each dairyman will decide on BST usage depending on how response is perceived rather than on discrete measurements of production response and financial advantage. Dairies best able to assess a BST advantage will record rolling 5-day-average milk weights throughout treatment, using farmer-owned milk-weighing devices linked to computers such as sold by major milking equipment companies.

Selecting Cows for BST Treatment

It is known that different herds and cows within a herd respond differently to various levels of BST supplementation (Crooker and Otterby 1990). Although it would be desirable to select cows that will respond maximally to BST in advance, at present there is no viable scheme by which to do so (Peel et al. 1989).

Some studies show that milk production at treatment initiation has little effect on the level of response to BST (Gibson et al. 1990, McDaniel and Hayes 1988). On the other hand, Chalupa et al. (1986) found that in Pennsylvania herds higher levels of production resulted in a greater BST response. Some of the differences in these findings are thought to be due to management factors, although evidence to support this premise was not reported.

Some researchers have found that genetic potential has an effect on BST response, but the direction and size of response has been variable among cows and reports. Nytes et al. (1990) found no effect of genetic potential on BST response. North Carolina workers (McDaniel et al. 1990a) found that cows of higher genetic potential responded better to BST than did their herdmates of lower genetic potential. Canadian research (Gibson et al. 1990) has shown just the reverse. That is, cows of lower genetic potential responded better to BST supplementation than did cows of higher genetic potential. The Canadian researchers postulated that this difference in production may be due to lower natural BST levels in the genetically inferior cows. Jersey, Holstein, Ayrshire, and Brown Swiss cows are known to respond at approximately the same level (Pell et al. 1988, Chalupa et al. 1987). In all studies, any effect of genetic potential or breed differences on BST response were small in comparison to the main BST effect. Genetic potential did not seem to be a significant factor in selecting cows for BST supplementation.

Numerous studies have shown that cows in midlactation (100 to 200 DIM) respond better to BST supplementation than do cows supplemented with BST earlier or later in lactation (Peel et al. 1989, Jordan et al. 1991). However, other studies have shown that starting BST treatment before 100 or after 200 DIM enhances milk yield at the same rate (McGuffey et al. 1991, Hansen et al. 1989, Cleale et al. 1989). The differences in response to time of initial treatment may relate to energy balance. Given the biologic action of BST, one would predict less of a response to BST supplementation for early-lactation cows in negative energy balance. While general management recommendations have been made for delaying treatment until 100 days into lactation (Thomas et al. 1991), perhaps a better recommendation would be to treat cows that are not experiencing weight loss. This would require accurate measures of weight or body condition scores for optimal management. Energy balance would be influenced by the production level as well as by days-in-milk (Patton 1989). Likewise, poor response in late lactation may be indicative of an overall hormonal balance that is not amenable to BST supplementation.

There is evidence (McDaniel et al. 1989, Phipps et al. 1990, McGuffey et al. 1991, Hansen et al. 1989) that BST may lengthen the calving interval by delaying conception. Thus, one current strategy is to delay use of BST until after confirmed pregnancy or at least until the cow has had several breedings. The delay in conception amounts to about one breeding cycle. On individual farms, it will require a good cost/benefit analysis program to determine if delaying BST administration until after confirmed pregnancy is a good management strategy. Field trial dairymen often put problem breeders on BST to maintain efficiency of production and extend the breeding window. They continued to breed these cows, but with less-expensive semen. In most cases, conception occurs while the cow is still producing at profitable levels, although the days-in-milk period may be extended. Salvaging these problem breeders may be cheaper than raising replacements, but it also delays genetic progress.

Initially, many popular-press articles suggested that body condition at the start of BST treatment would have a major effect on milk-production response. Because feed intake lagged behind production response, it was presumed that cows with greater body reserves could mobilize more nutrients for milk production under the stimulus of BST than could cows with less reserves. Controlled studies (Peel et al. 1989, Thomas et al. 1991), however, have not shown an effect of current body condition on BST response. On the other hand, long-term BST response and herd-production levels may be related to body condition at dryoff (Rijpkema et al. 1990). Field experience (Patton, unpublished) has found that unless cows are emaciated (BCS ~ 1), there is no correlation between body-condition score and milk-production response to BST. Cows with a body score of 2 or greater could be candidates for BST supplementation. Greater BST response is not obtained in fatter cows. However, poor body condition may impact subsequent lactations, particularly in cows in second lactation. More experience is required if BSC is to become a criterion for selecting cows for BST treatment. With or without BST use, cows must be managed to be in the proper body condition at dryoff.

The effect of mastitis on BST response has been studied. One study indicated a lesser BST response when pretreatment milk somatic cell count (SCC), an indirect measure of mastitis, is elevated (Thomas et al. 1991). Other studies have shown no relationship between pretreatment SCC and milk-production response (Hartnell et al. 1991). McDaniel et al. (1989) found that the cow's prior mastitis-treatment record was not a good predictor of BST response. Since BST increases the amount of milk synthesized per cell and increases mammary-cell longevity without apparently increasing mammary cell numbers (Knight et al. 1990), and given that mastitis infection reduces the number of cells actively secreting milk, it is reasonable to assume that mastitis-infection level would affect BST response. Although this assumption needs rigorous testing, there is enough evidence to suggest mastitis would decrease BST response. Therefore, cows with clinical mastitis are less likely to be financially profitable candidates for BST treatment. Further, it is reasonable to predict that those herds with lower rates of mastitis infection and subject to conscientious mastitis-control programs, would benefit most from BST use. Conversely, it does not appear that extraordinary measures of mastitis control will be required with BST-treated cows.

There are reports that BST use will raise herd SCC (Chapter 3). Other research has shown no effect (Chalupa et al. 1987, Eppard et al. 1987, Hartnell et al. 1991). Ferguson et al. (1990) and McClary et al. (1990) report that increased SCC are due entirely to increased milk production, although it is well known that higher-producing herds generally have lower SCC (Jones et al. 1984). At this point some somatic cell fundamentals need to be outlined. For details of this subject see Heald (1985) and Harmon and Heald (1989). The major source of somatic cells during mastitis are blood polymorphonuclear leukocytes. These are drawn from the blood into milk spaces by chemotactic agents today recognized as bacterial toxins and cytokines. Cytokines are agents produced by cells near the infection cite that enhance the response to foreign agents. Milk dilutes these chemotactic agents, toxins, and bacteria and washes away all agents to a degree at each milking. Thus increased milk flow is therapeutic and dilutes agents that contribute to elevated somatic cell counts. Increased milk production in cows generally lowers SCC by dilution. Considering the number of experiments that report a wide variance in SCC response to BST, along with the difficulty of conducting meaningful mastitis research in small clinical trials, it is probable that SCC increases are related to factors other than BST.

SCC data is highly accurate and inexpensive to collect, but is only an indirect measure of mastitis incidence and is related to severity. Bacterial cultures and clinical-treatment records are necessary for most complete assessment. Thus, reliance on SCC data alone to evaluate mastitis incidence in BST trials is not a sound procedure, and further research is required to understand the mechanism that leads to slight elevations in SCC. Such elevations are not likely to be significant since mastitis lowers milk yield and BST trials demonstrate increased yield. Explanations for high SCC other than BST use can be postulated but remain untested. Antidotal experience of one herd manager in field trials demonstrates how confusing evidence can be produced under field conditions. After initiation of BST treatment, 15 percent of the treated group experienced trampled teats and therefore clinical mastitis. These cows were all lower producers who had been assigned to smaller, less rigorously maintained stalls. The farmer speculated that smaller stalls had increased the incidence of teat injury and thus the treated group's mastitis rate.

Although no study has been done where diseased cows were intentionally given BST, consulting veterinarians have documented that in those herds where respiratory diseases have occurred during field trials (Wood and Simms 1991), production response to BST is considerably suppressed. Negative effects on the course of the recovery were not experienced. From an economic standpoint, owners of herds or cows with treatable diseases should probably delay BST use until the disease is corrected.

Considering what is known about BST response, it is not feasible to predict the response of individual cows or to select individual cows for BST supplementation. To date, the best strategy for maximizing the response to BST is to select groups of cows that meet the following general characteristics: healthy cows with positive energy balance, adequate dry-matter intake, acceptable body condition, and without clinical signs of mastitis. Production response should then be measured and the economic consequences calculated on this group. Cows with chronic disease that inhibit milk production (mastitis, respiratory, metabolic, leukosis, and foot and leg problems causing lameness) probably will experience reduced benefits from BST treatment.

Other Management Issues

Calf and Heifer Management

Studies have shown that BST treatment of dams produces no effect on body size, weight, or growth rate of the calf (Anderson et al. 1989, Hansen et al. 1989) even when the cow is treated late in pregnancy (Wilfond et al. 1989). Therefore, given our current knowledge of BST action, changes in calf management are not necessary. The best management and optimal nutrition of calves should be practiced for the productivity and profitability of the farm, regardless of BST usage.

The same can be said for management of growing heifers. However, heifer management is sometimes neglected on commercial dairies and may need greater emphasis to maximize BST response. Summarized data (Peel et al. 1989) indicates that with similar doses of BST, primiparous cows may have a lower production response than multiparous cows, perhaps because of lower dry-matter intake. This effect is not generalized among all herds, however (Thomas et al. 1991, de Boer and Kennelly 1989). Thomas and Patton (unpublished observations) have indicated that primiparous cows of adequate size (580 kg for Holsteins) and body condition (scoring 3.5-4.0) respond to BST supplementation at levels similar to multiparous cows. This shows that for optimum return on BST use, greater emphasis must be placed on growing heifers of adequate size, weight, and body condition. At present, there is no evidence that growing heifers need to be managed differently from the currently recommended practices. The evidence favors those dairy farms which have conscientious heifer-raising programs in place and are considering BST use.

Body-condition Management

Effects of BST supplementation on body condition score (BCS) vary (Chilliard 1989). Some studies (Samuels 1988, Arambel et al. 1989) have shown no effect on BCS, but others (McDaniel et al. 1990b, Thomas et al. 1991) have demonstrated a transitory negative effect.

Most studies, however, showed that regaining optimal body condition takes longer in BST-treated cows than in control animals (Phipps et al. 1990, McGuffey et al. 1991, Thomas et al. 1989). Some of these observed differences may be related to the length of the studies, the inaccuracy of the body-scoring technique, the rations fed, and the fact that BST-treated cows have more body condition to regain. Generally, cows must be on BST for several months before changes in BCS appear. Considering that one of the known actions of BST is to decrease lipogenesis (production of body fat), it should be expected that replacing body condition may require careful feed management in BST-treated cows as well as in similarly high-producing cows not treated with BST.

Although there are no studies which demonstrate an effect of body condition on BST response in the initial-treatment lactation, failure to regain adequate body reserves at dryoff has suppressed subsequent lactation performance (Gibson et al. 1990) and BST response (Phipps et al. 1990). Body condition affects production (Seymour and Polan 1986) and reproduction (Ferguson 1989) in non-BST-treated cows. Also, high-producing herds have more difficulty in maintaining adequate body condition, and the lack of body condition may cap the production of some herds (Erdman et al. 1989).

From a management perspective, this suggests that management of body condition will become increasingly important with higher production and widespread BST use. Some authors (Chalupa and Galligan 1989, Palmquist 1988) have suggested the use of high-fat diets or diets high in ruminally bypass fat to restore body condition. Thomas et al. (1990) reported Michigan field trial herds with which they had worked experienced normal body-condition replenishment because these farmers practiced "good feed management."

Once approved, it is expected that BST will be used on farms with diverse management styles and facilities that can effect BST response. Bunk space, ventilation, duration of feed availability, shade, water quality, feed palatability, quality of silage fermentation, and other factors will cause variation in milk-yield response, body condition, weight gain, and farm profit. Additional field experience with high-producing cows will improve the industry's knowledge of when such factors are important and to what degree. Until this fine-tuning is better understood, dairy farmers will optimize BST response if the following strategy is used:

1. Feed diets based on milk production and body condition score such that cows attain proper body condition at dryoff: in other words, feed complete diets *ad libitum*.
2. Optimize dry-matter intake.
3. Terminate BST use early for cows in poor condition.
4. As a last resort, modify dry-cow feeding to encourage weight gain if body condition cannot be regained fully during the latter part of lactation.

As with all studies to increase milk production, additional studies are needed to confirm these observations and to develop specific recommendations for managing body condition in dairy cows at high levels of production.

Feeding Management

Kalter et al. (1984), in an early study of the impact of BST on dairy farm profitability, suggested that farmers with computerized automatic feeders would profit most from BST usage. Controlled studies to confirm this speculation are not available. However, Thomas et al. (1987) found that milk-yield response was similar in BST-treated herds where cows were fed with a component system or total mixed rations (TMR). In this experiment, BST-treated cows fed with the component system gained less weight. A limited number of studies have investigated various feeding schemes, including component feeding (Pell et al. 1988), flat-rate feeding (Furniss et al. 1988), magnet feeders (Thomas et al. 1990), and parlor feeders (Thomas et al. 1991). Milk-production responses in each case was approximately 5 kg/cow/day. There is some variation in the data, but no apparent difference attributable to feeding system. However, feeding grain separate from other feedstuffs meals should be scrutinized when practiced. Done in excess, it may interfere with rumen fermentation or induce an insulin surge that can interfere with BST action (Chase et al. 1977).

The nutrition of BST-supplemented cows is reviewed in Chapter 3. Authors have proposed various schemes for feeding the BST-supplemented cow: increasing the amount of grain fed (Tessman et al. 1988); keeping grain constant and feeding high-quality forage *ad libitum*; feeding a single, high-energy TMR (Hutjens 1990); or increasing the amount of ruminally bypassed fat (Chalupa and Galligan

1989) and protein (McGuffey et al. 1991). Although many workers support these strategies, little research exists to adequately substantiate these recommendations. French et al. (1990) found no effect of feeding frequency on BST response. At present, encouraging increased DMI is a proven feeding strategy (Franson et al. 1989, Tyrrell et al. 1988). Based on these observations and on personal experience in the field, feeding a continuously available TMR balanced at a suitable level of production may be the most productive method of feeding BST-treated cows. Equipment to deliver more concentrates more times per day or the development of complicated rations do not appear necessary for satisfactory BST results. The advantage of high-quality forages for maximizing DMI should also be emphasized.

Observation and some research (Thomas et al. 1991) show that feeding management may play a greater role than ration balance in BST response, at least in the short term. Herd-production level appears to be a reflection of both ration balance and DMI. Factors that increase DMI, such as optimum ration balance, particular hours of the day when feed is available, adequate bunk space, separate grouping of primiparous and multiparous cows, controlling competition between aggressive and timid cows, increased frequency of feeding, controlling holding-pen group size, and managing heat stress with proper ventilation, shades, and sprinklers should be optimized within each farm operation (Ferry 1989). These are strategies that have been observed to increase DMI and BST response in field trials. These are also important for herds that do not use BST.

BST field trials conducted in Pennsylvania illustrate another point. Response to BST appeared to be more dependent on the hours of feed availability than on herd-production average (Table 4.1). The one instance in which this was not the case (Herd F) was an instance in which extremely poor forage was offered about midway through the experiment. This resulted in prolonged periods of decreased DMI. It appears that those managers who pay close attention to these aspects of their feeding systems and adjust them to obtain maximum DMI will be the ones who experience the greatest BST response.

Upon the commercial introduction of BST, there will be many opportunities to study the effects of different feeding strategies in a variety of settings. It will be of great benefit to dairy farmers if industry advisors and Extension personnel document these experiences and reduce them to practical recommendations as soon as possible.

TABLE 4.1. Rolling Herd Average, BST Response, and Hours of Feed Availability for Six Pennsylvania Field Trial Herds.

Herd	*Rolling Herd Average (kg/cow/year)*	*BST Response (kg/day)*	*Feed Available (hour/day)*
A	7,905	3.9	14
B	8,166	4.3	16
C	8,182	5.1	19
D	6,756	6.9	22
E	9,027	8.6	22
F	7,584	2.1	22

Forage Quality

Sniffen (1989) and Hutjens (1990) suggested that those dairymen who feed high-quality forage will see a greater BST response than those feeding lower-quality forage. In addition to the well-documented general effects of high-quality forage on milk production, BST field trials in Idaho and Pennsylvania support these assertions.

Two trial herds in Idaho, both housed, fed, and managed in similar fashion, were offered, free-choice hay of different quality (32% vs 47% acid detergent fiber [ADF]). The herd fed the lower-quality hay had about a 2 kg milk/cow/ day lower BST response (Lamb, unpublished). More enlightening is a study of one herd in Pennsylvania, also unpublished. This herd employed a component feeding system, and cows were housed in con-ventional tie stalls. BST response was adequate until the herd was switched to a low-quality haylage. Following the change, BST response declined dramatically (Figure 4.1). It is speculated that this apparent apparent effect is due to reduced DMI and therefore, lower energy consumption. Unknown factors could have coincidentally caused these differences in response, lower forage quality could reflect an overall lower level of management. However, it does suggest that forage quality should be researched as a predictor of BST response.

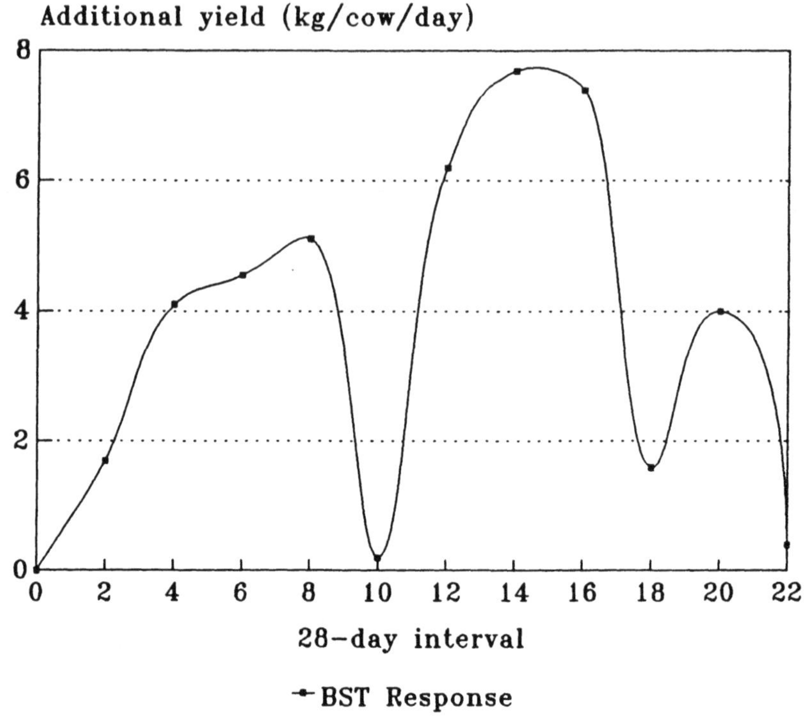

Source: Table 4.1 (herd F).

FIGURE 4.1. Daily response (kg/cow) in milk yield in a Pennsylvania field trial with BST supplementation.

Dry Cows

Management of dry cows is a separate, but related, issue that impacts BST use. The major concerns are body condition at dryoff, incidence of peripartum disease, and duration of negative energy balance during the previous lactation. Some articles in the popular press have suggested that somatotropin-treated cows be put on high-energy diets at dryoff to regain body condition during the dry period. Little published work exists to confirm or reject this suggestion. It is known that rapid weight gain in the dry period has been associated with increased incidence of peripartum disease (Emery and Herdt 1991). Peripartum disease, especially with liver involvement, has the potential to decrease BST response if this response is mediated by liver IGF-1 (Peel and Bauman 1987). Further it is more efficient to recover body stores during late lactation than in the dry period (Moe et al. 1971). This could be due to a reduced DMI and/or changes in hormonal balance. At present, there is little evidence to support departure from normal dry-cow management. To develop useful strategies, controlled dry-cow management studies are required over several lactations of BST use. Our current knowledge suggests that the best recommendation for BST cows is: cows should probably be managed to gain no more than one BCS (Reid and Collins 1980), even if body condition is poorer than desirable at dryoff. Dry cows losing condition is associated with the highest incidence of postpartum disease (Emery and Herdt 1991). Although no definitive BST studies exist, inadequate dry-cow nutrition, which predisposes a cow to peripartum disease and results in body condition loss, may explain some of the variation in BST response. Delaying use of BST until later in the lactation may be prudent in such cases.

If body condition is not regained during late lactation a less desirable alternative is to use two dry-cow feeding groups to gain as much body weight as possible without precipitating peripartum disease. Cows would be assigned to one of the following diets:

1. maintain body condition, and
2. gain body weight slowly.

Near parturition both groups could be fed a transition diet. Experience with this approach needs to be documented.

Managing the Calving Interval

Some studies (see Table 3.3 in Chapter 3) have shown a negative effect of BST on reproductive efficiency, resulting in an increased number of days open. Other workers (Chalupa and Galligan 1989, Ferguson 1989, Samuels et al. 1988) have reported that longer-term use of BST fundamentally changes the shape of the lactation curve such that BST-treated cows may not experience the milk-production decline associated with normal lactation. Also, more time following peak production may be required for BST-supplemented cows to replenish body condition. Thus, scientists at Cornell suggest that calving interval may need to be lengthened with BST use. Optimum calving intervals vary from farm to farm and depend on the involuntary culling rate (Allaire et al. 1977, Dentine and McDaniel 1987, Rogers et al. 1988b), the reason for culling, the production level at which farmers choose to cull, the age at first freshening, the cost of feed, and the seasonality of milk price. In North Carolina studies over three years, culling rate did not change appreciably with BST use (McDaniel et al. 1990a). However, BST-treated cows and nontreated cows were culled for different reasons. BST-treated cows were less-heavily culled for low production, but tended to be subjected to increased culling pressure for mastitis and reproductive problems. BST-treated cows remained in the herd the same length of time but at higher levels of production. Some authors (Ferguson 1989, Ferguson et al. 1990) have suggested that increased mastitis and reproductive problems are entirely explained by increased milk production and increased age. European studies of longer term (Weller et al. 1990), however, have not shown differences in either culling rate or reasons for culling. These results may be different because of the culling strategies employed which in turn are reflective of the milk-market conditions on the two continents. BST field studies with high-producing herds in Michigan (Thomas et al. 1990) have shown minor declines in culling rate with only slight changes in calving interval, an indication that better management of farm practices minimizes mastitis and inefficiency of reproduction.

On the basis of published research, there seems to be no compelling reason to change calving interval. However, if due to BST-supplementation, herd production remains high enough to be profitable, or if body condition remains poor, or if replacement animals are not available, then increasing the calving interval may be an acceptable management strategy to maintain overall herd production. There may also be benefits to reducing the number of overall lifetime calvings and, therefore, the health risks associated with parturition and dry-cow mastitis. Determination of ideal calving interval with BST use must await wide adoption of BST and more refined computer models before calving interval recommendations can be made.

Cow Management

Some dairymen have asked if it will be necessary to increase milking frequency if BST is used. BST has been used on high-producing herds milked two (Bauman et al. 1985), three (Jordan et al.

1991) and four times per day (Armstrong et al. 1990). In all studies, the effects of BST and milking frequency were found to be additive. Desired herd-production level, milk price, and labor availability, rather than BST use, appear to determine the desirability of milking more frequently than two times per day.

Cows managed at high environmental temperatures are expected to be less responsive to BST (Staples et al. 1988, West et al. 1990). Indeed, BST supplemented cows have slightly higher rectal temperatures and increased respiration rate (Johnson et al. 1991, Zoa-Mboe et al. 1989, Mohammed and Johnson 1985). Elevated temperatures were associated with decreased milk production, but not with a decreased BST response. In Arizona studies (Armstrong et al. 1990 and Duque et al. 1990), daily high temperatures varied between 38 and 48°C and BST response across several herds was reasonably high (~5 kg/cow/day). However, studies at lower ambient temperature, but presumably higher humidity, in Florida (Staples et al. 1988) and in South Carolina (Jenny et al. 1989), found a reduced BST response (~2 kg/cow/day). It might be hypothesized that humidity had a greater impact than temperature on BST response. Critical studies to confirm this postulate are absent. It is important to determine the environmental conditions and milk price at which BST treatment ceases to be profitable. Until this research is performed, management strategies for effective BST use in hot, humid climates should focus on minimizing heat stress--provide adequate shade, evaporative cooling, supplemental ventilation, adequate water, and maximize dry-matter intake.

Several observations suggest that cow comfort may have an influence on BST response, but little research has been done. Field-trial dairy farmers perceived in times of stress that BST-supplemented cows decrease production more than do their untreated herdmates, although absolute yields remained higher. These dairy farmers reasoned that any cow producing more milk needs to eat, rest, and ruminate more. Therefore, any condition that would encourage these activities should result in greater response, and deterrents to these activities should be removed. To this end, tie stalls and free stalls were redesigned and kept continually clean and bedded. Barn walls were removed to improve ventilation. Feeders were redesigned and clean drinking water was made more available. Barn yards were cleaned and muddy areas eliminated. Herd-production levels and BST-response improved as cow comfort improved. New management techniques were not required to elicit more response, only application of present-day recommended practices. Such changes can enhance breeding efficiency and reduce mastitis incidence.

Little is known about which cow-comfort factors optimize BST response and how they may be manipulated. This should prove a fruitful area for future research.

BST Use as a Milk Marketing Tool

Ultimately the decision to use BST should be based on economic return or advantages in production management. It is possible that BST use in some herds will be justified during certain times of the year or under certain economic conditions. Perhaps BST use with only select cows within the herd will be economically beneficial. Enhanced milk yield from BST use offers managers the option to keep cows open longer, or to keep "problem breeders" in the herd longer while attempting to achieve pregnancy. BST use could also increase cash flow in seasons of tight supply, an alternative to adding more cows or retaining marginally profitable cows, as is often now the case. In areas where environmental regulations limit the number of cows that can be housed on a given farm, BST may offer the manager an opportunity to increase production without adding substantially to the environmental problem. The same may be true for dairymen near urbanized areas.

Producers in marketing areas where a quota system is in place and herd production is below quota could elect to use BST as a means of achieving quota without additional cows. Conversely, if production is over quota, producers could elect to reduce BST use or elect to reduce cow numbers. For those producers that market under some type of seasonal-incentive plan (Louisville or Base-Excess Plan), BST use may provide a tool to maximize the price differentials. For example, at base-building time or at time of greatest pay back, a farmer may want to treat a higher percentage of his herd with BST in order to maximize his base shippings. Consequently, he may wish to reduce the percent treated, or sell the least profitable cows, in order to minimize his over-base shipping at other times of the year. Additionally, in areas with high fluid utilization, dairymen may wish to use BST to maximize shipping when weather conditions are less conducive to high production. Such a scenario has been proposed for southeastern dairy farmers (Jenny 1990), where demand for milk in the fall of the year far out-strips farmers' ability to produce it. Such a strategy could well depend on other factors, such as dry-cow management, forage quality, facilities to minimize heat stress, and feed availability. Others may consider use of BST in the fall of the year to recover more quickly from summer slump. Presumably, many dairy herds, even those with less than optimum management, could make some use of BST for a 3- to 5-month period as a means of maximizing returns from marketed milk.

Conclusion

Although there is general recognition that cow management plays a role in BST use and response, little controlled research is available as to the management adjustments that must be made to optimize BST response. The interaction of BST with management, and development of optimal BST-management systems promises to be a fruitful area of research. The widespread use of BST could accelerate technology adoption currently available from Extension, industry, and consultants. At present, progressive dairy farmers employing "best-management" strategies (e.g., balancing rations, maximizing DMI, following mastitis-control program and production-medicine programs, feeding high-quality forages, and advanced production-monitoring and farm-cost accounting records), appear to have the greatest potential to optimize profits from BST use. After more research, adjustments may be made in the management of the feeding program, and management of body reserves to optimize profitable production response. Based on current knowledge of BST and its effects on dairy cow performance, our best advice for users of BST is to follow the practices that have proven effective in managing high-producing herds before recombinant-BST was available.

The use of BST will further complicate farm-management options. BST use may not always be profitable. Herds at high levels of production must be monitored closely and any negative situation corrected early if the high production level is to be maintained. Careful financial and production records will be even more necessary on present and future dairy farms.

BST use also appears to allow farmers more latitude in managing herd production for regional milk markets. Transient farm alterations in production patterns could also be smoothed out by use of BST. Significant challenges remain for the dairy industry if it is to develop information systems and consulting services that will allow farmers to manage BST and other technologies for maximal profit.

Acknowledgments

Dairy farmers who have contributed to our understanding of the role of cow comfort in BST response include: Ken and Larry Nobis, Michigan; Larry and Wayne Webster, Michigan; Doug Young, New York; Jon Merrill, New York; John Piwowar, Pennsylvania; and John and Jonas Esh, Pennsylvania.

Many individuals have provided suggestions leading to improvements to this manuscript, but special thanks go to Dr. G. M. Jones of Virginia Polytechnic Institute and State University for his objective evaluation and critical review which strengthened the clarity and readability of this chapter.

References

Allaire, F. R., H. E. Sterwerf, and T. M. Ludwick. 1977. "Variations in removed reasons and culling rates with age for dairy females." *Journal of Dairy Science* 60:254-267.

Anderson, M. J., R. C. Lamb, R. J. Callan, G. F. Hartnell, R. G. Hoffman, L. Kung, Jr., and S. E. Franson. 1989. "Effect of sometribove (recombinant methionyl bovine somatotropin) on gestation length and on body measurements, growth, and blood chemistries of calves whose dams were treated with sometribove." *Journal of Dairy Science* 72(Suppl):327. (Abstract).

Arambel, M. J., S. O. Oellerman, R. C. Lamb, and G. A. Green. 1989. "The effect of sometribove (recombinant methionyl bovine somatotropin) on production responses in lactating dairy cows: A field trial." *Journal of Dairy Science* 72(Suppl):451. (Abstract).

Armstrong, D. V., A. Burgos, J. A. Duque, and K. S. Madsen. 1990. "The effect of sometribove (recombinant methionyl bovine somatotropin) on milk yield in lactating dairy cows milked 4 times a day in a commercial dairy herd." *Journal of Dairy Science* 73(Suppl 1):160. (Abstract).

Bauman, D. E. 1989. "Biology of bovine somatotropin in dairy cattle." Pp 1-8 in *Advanced Technologies Facing the Dairy Industry: BST*. Ithaca, NY: Cornell Cooperative Extension Animal Science Mimeograph Series #133.

Bauman, D. E., P. J. Eppard, M. J. DeGeeter, and G. M. Lanza. 1985. "Responses of high producing cows to long-term treatment with pituitary somatotropin and recombinant somatotropin." *Journal of Dairy Science* 68:1352-1362.

Bauman, D. E., D. L. Hard, B. A. Crooker, M. S. Partridge, K. Garrick, L. D. Sandles, H. N. Erb, S. E. Franson, G. F. Hartnell, and R. L. Hintz. 1989. "Long-term evaluation of a prolonged-release formulation of N-methionyl bovine somatotropin in lactating dairy cows." *Journal of Dairy Science* 72:642-651.

Chalupa, W., L. Baird, C. Soderholm, D. L. Palmquist, R. Hemken, D. Otterby, R. Annexstad, B. Vecchiarelli, R. Harmon, A. Sinha, J. Linn, W. Hansen, F. Ehle, P. Schneider, and R. Eggert. 1987. "Responses of dairy cows to somatotropin." *Journal of Dairy Science* 70(Suppl):176. (Abstract).

Chalupa, W., and D. T. Galligan. 1989. "Nutritional implications of somatotropin for lactating cows." *Journal of Dairy Science* 72:2510-2524.

Chalupa, W., B. Vecchiarelli, P. Schneider, and R. G. Eggert. 1986. "Long-term responses of lactating cows to daily injection of recombinant somatotropin." *Journal of Dairy Science* 69(Suppl):151. (Abstract).

Chase, L. E., P. J. Wangsness, J. F. Kavanaugh, L. C. Griel, Jr., and J. H. Gahagan. 1977. "Changes in portal blood metabolites and insulin with feeding steers twice daily." *Journal of Dairy Science* 60:403-415.

Chilliard, Y. 1989. "Long-term effects of recombinant bovine somatotropin (rBST) on dairy cow performances: A review." Pp 61-87 in K. Sejrsen, M. Vestergaard, and A. Neimann-Sorensen (eds). *Use of Somatotropin in Livestock Production*. New York, NY: Elsevier Applied Science.

Cleale, R. M., J. D. Rehman, E. J. Robb, A. Sinha, F. R. Ehle, and D. K. Nelson. 1989. "On-farm lactational and reproductive responses to daily injections of recombinant bovine somatotropin." *Journal of Dairy Science* 72(Suppl):429. (Abstract).

Crooker, B. A. and D. E. Otterby. 1990. "Managing the BST treated cow." Pp 59-83 in *Proceedings of 51st Minnesota Nutrition Conference*. Bloomington, MN: University of Minnesota.

de Boer, G., and J. J. Kennelly. 1989. "Sustained release somatotropin for dairy cows." *Journal of Dairy Science* 72(Suppl):432. (Abstract).

Dentine, M. R., and B. T. McDaniel. 1987. "Comparison of culling rates, reason for disposal and yields for registered and grade Holstein cattle." *Journal of Dairy Science* 70:2616-2622.

Duque, J. A., K. S. Madsen, D. V. Armstrong, and A. Burgos. 1990. "The effect of sometribove (recombinant methionyl bovine somatotropin) on the milk response in lactating Jersey cows in a commercial dairy herd." *Journal of Dairy Science* 73(Suppl):160. (Abstract).

Erdman, R. A., and S. M. Andrew. 1989. "Methods for and estimates of body tissue mobilization in the lactating dairy cow." Pp 19-26 in *Meeting the challenges of New Technology*. Monsanto Technical Symposium, 24 Oct 1989, Syracuse, NY. St. Louis, MO: Monsanto Agricultural Company.

Erdman, R. A., L. W. Douglass, and M. A. Varner. 1990. "A dose response summarization model of published somatotropin research." *Journal of Dairy Science* 73(Suppl):162. (Abstract).

Emery, R. S., and T. H. Herdt. 1991. "Lipid nutrition." *Veterinary Clinic of North America* 7:341-352.

Eppard, P. J., D. E. Bauman, C. P. Curtis, and H. N. Erb. 1987. "Effect of 188-day treatment with somatotropin on health and reproductive performance of dairy cows." *Journal of Dairy Science* 70:582-591.

Ferguson, J. D. 1989. "Interactions between milk yield and reproduction in dairy cows." Pp 35-44 in *Meeting the Challenges of New Technology*. St. Louis, MO: Monsanto Technical Symposium, 24 Oct 1989, Syracuse, NY. Monsanto Agricultural Company.

Ferguson, J. D., D. Galton, and A. Skidmore. 1990. "Factors influencing the response to sometribove in commercial dairy herds." Ithaca, NY: Cornell University, Department of Animal Science.

Ferry, J. W. 1989. "Managing the BST treated herd--a veterinarian perspective." Pp 45-54 in *Meeting the Challenges of New Technology*. Monsanto Technical Symposium, 24 Oct 1989, Syracuse, NY. St. Louis, MO: Monsanto Agricultural Co.

Franson, S. E., W. S. Cole, R. G. Hoffman, V. K. Meserole, D. M. Sprick, K. S. Madsen, G. F. Hartnell, D. E. Bauman, H. H. Head, J. T. Huber, and R. C. Lamb. 1989. "Response of cows throughout lactation to sometribove, recombinant methionyl somatotropin, in a prolonged release system--a dose titration study." *Journal of Dairy Science* 72(Suppl):451. (Abstract).

French, N., G. de Boer, and J. J. Kennelly. 1990. "Effects of feeding frequency and exogenous somatotropin on lipolysis, hormone profiles and milk production in dairy cows." *Journal of Dairy Science* 73:1552-1559.

Furniss, S. J., A. J. Stroud, A. C. G. Brown, and G. Smith. 1988. "Milk production, food intakes and weight changes of autumn-calving, flat-rate-fed dairy cows given 2-weekly injections of recombinantly-derived bovine somatotropin (BST)." *Animal Production* 46:483. (Abstract).

Galton, D. M. 1989. "Evaluation of Sometribove, USAN (recombinant methionyl bovine somatotropin) on milk production and health." *Journal of Dairy Science* 72(Suppl):450. (Abstract).

Gibson, J. P., M. van der Meulen, B. W. McBride, and J. H. Burton. 1990. "The relationship between phenotypic and genetic milk yield potential and responses to rBST in lactating dairy cows. *Journal of Dairy Science* 73(Suppl):158. (Abstract).

Hansen, W. P., D. E. Otterby, J. G. Linn, J. F. Anderson, and R. G. Eggert. 1989. "Multi-farm use of bovine somatotropin (BST) and its effects on lactation, health and reproduction." *Journal of Dairy Science* 72(Suppl):429. (Abstract).

Hart, I. C. 1988. "Altering the efficiency of milk production of dairy cows with somatotropin." Pp 232-247 in P. C. Garnsworthy (ed). *Nutrition and Lactation in the Dairy Cow*. London, UK: Butterworths.

Harmon, R. J., and C. W. Heald. 1982. "The migration of polymorphonuclear leukocytes into the bovine mammary gland during experimentally induced Staphylococcus aureus mastitis. *American Journal of Veterinary Research* 43:992-998.

Hartnell, G. F., S. E. Franson, D. E. Bauman, H. H. Head, J. T. Huber, R. C. Lamb, K. S. Madsen, W. J. Cole, and R. L. Hintz. 1991. "Long-term evaluation of sometribove, recombinant methionyl bovine somatotropin, in a prolonged release system in lactating dairy cows - production responses in a dose titration study." *Journal of Dairy Science* 74:2645-2663.

Heald, C. W. 1985. "Milk Collection" Pp 198-228 in B. L. Larson (ed). *Lactation*. Ames, IA: The Iowa State University Press.

Hutjens, M. F. 1990. Nutritional Management and Economical Aspects of BST. Pp 13-19 in *Symposium of 15th Annual Food Animal Medicine Conference*. October 16-17, 1989, Columbus, Ohio.

Jenny, B. F. 1990. Personal Communication.

Jenny, B. F., M. Moore, R. B. Tingle, J. E. Ellers, L. W. Grimes, and D. W. Rock. 1989. "Effect of sustained release somatotropin on lactation performance of dairy cattle." *Journal of Dairy Science* 72(Suppl):431. (Abstract).

Johnson, H. D., R. Li, W. Manalu, K. J. Spencer-Johnson, B. Ann Becker, R. J. Collier, and C. A. Baile. 1991. "Effects of somatotropin on milk yield and physiological responses during summer farm and hot laboratory conditions." *Journal of Dairy Science* 74:1250-1262.

Jones, G. M., R. E. Pearson, G. A. Clabaugh, and C. W. Heald. 1984. "Relationships between somatic cell counts and milk production" *Journal of Dairy Science* 67:1823-1831.

Jordan, D. C., A. A. Aguilar, J. D. Olson, C. Bailey, G. F. Hartnell, and K. S. Madsen. 1991. "Effects of recombinant methionyl bovine somatotropin (sometribove) in high producing cows milked three times daily." *Journal of Dairy Science* 74:220-226.

Kalter, R. J. 1984. "Biotechnology and the dairy industry: production costs and commercial potential of the bovine growth hormone." AER-84-22. Ithaca, NY: Department of Agricultural Economics, Cornell University.

Knight, C. H., P. A. Fowler, and C. J. Wilde. 1990. "Galactopoietic and mammogenic effects of long term treatment with bovine growth hormone and thrice daily milking in goats." *Journal of Endocrinology* 127:129-138.

McBride, B. W., J. L. Burton, and J. H. Burton. 1988. "The influence of bovine growth hormone (somatotropin) on animals and their products." *Research Devevelopment in Agriculture* 5:1-21.

McClary, D., R. K. McGuffey, and H. B. Green. 1990. "Bovine somatotropin: Part 2." *Agri-Practice* 11:13-16.

McDaniel, B. T., and P. W. Hayes. 1988. "Absence of interaction of merit for milk with recombinant bovine somatotropin." *Journal of Dairy Science* 71(Suppl):240. (Abstract).

McDaniel, B. T., W. E. Bell, J. Fetrow, B. D. Harrington, and J. D. Rehman. 1990b. "Survival rates and reasons for removal of cows injected with rBST." *Journal of Dairy Science* 73(Suppl):159. (Abstract).

McDaniel, B. T., J. Fetrow, B. D. Harrington, W. E. Bell, and J. D. Rehman. 1990a. "Factors affecting response to recombinant bovine somatotropin." *Journal of Dairy Science* 73(Suppl):159. (Abstract).

McDaniel, B. T., D. M. Gallant, J. Fetrow, B. Harrington, W. E. Bell, P. Hayes, and J. D. Rehman. 1989. "Lactational, reproductive and health responses to recombinant somatotropin under field conditions." *Journal of Dairy Science* 72(Suppl):429. (Abstract).

McGuffey, R. K., R. P. Basson, D. L. Snyder, E. Block, J. H. Harrison, A. H. Rakes, R. S. Emery, and L. D. Muller. 1991. "Effects of somidobove sustained release administration on the lactation performance of dairy cows." *Journal of Dairy Science* 74:1263-1276.

McGuffey, R. K., H. B. Green, and R. P. Basson. 1990. "Lactation response of dairy cows receiving bovine somatotropin and fed rations varying in crude protein and undegradable intake protein." *Journal of Dairy Science* 73:2437-2443.

Marsh, W. E., D. T. Galligan, and W. Chalupa. 1988. "Economics of recombinant bovine somatotropin use in individual dairy herds." *Journal of Dairy Science* 72:2944-2958.

Mix, L. S. 1987. "Potential impact of the growth hormone and other technology on the United States dairy industry by the year 2000." *Journal of Dairy Science* 70:487-497.

Moe, P. W., H. F. Tyrrell, and W. P. Flatt. 1971. "Energetics of body tissue mobilization." *Journal of Dairy Science* 54:548-553.

Mohammed, M. E., and H. D. Johnson. 1985. "Effect of growth hormone on milk yields and related physiological functions of Holsteins exposed to heat stress." *Journal of Dairy Science* 68:1123-1133.

Nytes, A. J., O. K. Combs, G. E. Shook, R. D. Shaver, and R. M. Cleale. 1990. "Response to recombinant bovine somatotropin in dairy cows with different genetic merit for milk production." *Journal of Dairy Science* 73:784-791.

Palmquist, D. L. 1988. "Response of high producing cows given daily injection of recombinant bovine somatotropin from d 30-296 of lactation." *Journal of Dairy Science* 72(Suppl):206. (Abstract).

Patton, R. A. 1989. "The effect of dietary fiber and body condition on the milk production, dry-matter intake and blood metabolites of peripartum dairy cows." Phd. Dissertation, Michigan State University, East Lansing, MI.

Peel, C. J., and D. E. Bauman. 1987. "Somatotropin and lactation." *Journal of Dairy Science* 70:474-486.

Peel, C. J., D. L. Hard, K. S. Madsen, and G. de Kerchove. 1989. "Bovine somatotropin: mechanism of action and experimental results from different world areas." Pp 9-18 in *Meeting the Challenge of New Technology*. Monsanto Technical Symposium preceding the Cornell Nutrition Conference. St. Louis, MO: Monsanto Agricultural Co.

Pell, A. N., D. S. Tsang, M. T. Huyler, B. A. Howlett, J. Kunkel, and W. A. Samuels. 1988. "Responses of Jersey cows to treatment with sometribove, USAN (recombinant methionyl bovine somatotropin) in a prolonged-release system." *Journal of Dairy Science* 71(Suppl):200. (Abstract).

Phipps, R. H., R. F. Weller, N. Craven, and C. J. Peel. 1990. "Use of prolonged-release bovine somatotropin for milk production in British Freisian dairy cows. I. Effect on intake, milk production and feed efficiency in two consecutive lactations of treatment." *Journal of Agriculture Science* 115:94-104.

Prosser, C. G., I. R. Fleet, A. N. Corps, E. R. Froesch, and R. B. Hemp. 1990. "Increase in milk secretion and mammary blood flow by intra-arterial infusion of insulin-like growth factor-I into the mammary gland of the goat." *Journal of Endocrinology* 126:437-443.

Reid, I. M., and R. A. Collins. 1980. "The pathology of post-parturient fatty liver in high yielding dairy cows." *Investigative Cellular Pathology* 3:237-249.

Reid, I. M., C. J. Roberts, R. J. Treacher, and L. A. Williams. 1986. "Effect of body condition at calving on tissue mobilization, development of fatty liver and blood chemistry of dairy cows." *Animal Production* 43:7-15.

Rijpkema, Y. S., L. van Reeluwijk, and D. L. Hard. 1990. "Response of dairy cows to sometribove (r-BST) during three consecutive years." *Livestock Production Science* 26:193-216.

Rogers, G. W., J. A. M. Van Arendonk, and B. T. McDaniel. 1988a. "Influence of production and prices on optimum culling rates and annualized net revenue." *Journal of Dairy Science* 71:3453-3462.

Rogers, G. W., J. A. M. Van Arendonk, and B. T. McDaniel. 1988b. "Influence of involuntary culling on optimum culling rates and annualized net revenue." *Journal of Dairy Science* 71:3463-3469.

Samuels, W. A., D. L. Hard, R. L. Hintz, P. K. Olsson, W. J. Cole, and G. F. Hartnell. 1988. "Long term evaluation of sometribove, USAN (recombinant methionyl bovine somatotropin) in a prolonged release system for lactating cows." *Journal of Dairy Science* 71(Suppl):209. (Abstract).

Seymour, W. M. and C. E. Polan. 1986. "Dietary energy regulation during gestation on subsequent lactational response to soybean meal or dried brewers grain." *Journal of Dairy Science* 69:2837-2845.

Sniffen, C. J., W. Chalupa, and J. Ferguson. 1989. "The impact of controlling protein, amino acid and carbohydrate fractions on productivity and body weight change in BST herds." Pp 27-33 in *Meeting the Challenges of New Technology*. Monsanto Technical Symposium, 24 Oct 1989, Syracuse, NY. St. Louis, MO: Monsanto Agricultural Co.

Staples, C. R., H. H. Head, and D. E. Darden. 1988. "Short-term administration of bovine somatotropin to lactating cows in a subtropical environment." *Journal of Dairy Science* 71:3274-3282.

Tessman, N. J., J. Kleinmans, T. R. Dhiman, H. D. Radloff, and L. D. Satter. 1988. "Effect of dietary forage:grain ratio on response of lactating dairy cows to recombinant somatotropin." *Journal of Dairy Science* 71(Suppl):121. (Abstract).

Thomas, C., I. D. Johnsson, W. J. Fisher, G. A. Bloomfield, S. V. Morant, and J. M. Wilkerson. 1987. "Effect of somatotropin on milk production, reproduction and health of dairy cows." *Journal of Dairy Science* 69(Suppl):175. (Abstract).

Thomas, J. W., R. A. Erdman, D. M. Galton, R. C. Lamb, M. J. Arambel, J. D. Olson, K. S. Madsen, W. A. Samuels, C. J. Peel, and G. A. Green. 1991. "Responses by lactating cows in commercial dairy herds to recombinant bovine somatotropin." *Journal of Dairy Science* 74:945-964.

Thomas, J. W., W. A. Samuels, and C. J. Peel. 1990. "Evaluation of management practices and sometribove, USAN (recombinant methionyl) bovine somatotropin usage in Michigan dairy herds during two years." *Journal of Dairy Science* 73(Suppl):158. (Abstract).

Thomas, J. W., W. A. Samuels, and K. S. Madsen. 1989. "Use of sometribove, USAN (recombinant methionyl) in a prolonged release system in commercial dairy herds." *Journal of Dairy Science* 72(Suppl):450. (Abstract).

Tyrrell, H. F., A. C. G. Brown, P. J. Reynolds, G. C. Harland, D. E. Bauman, C. J. Peel, and W. D. Steinhour. 1988. "Effect of bovine somatotropin on metabolism of lactating dairy cows: Energy and nitrogen utilization as determined by respiration calorimetry." *Journal of Nutrition* 118:1024-1030.

Weller, R. F., R. H. Phipps, N. Craven, and C. J. Peel. 1990. "Use of prolonged-release bovine somatotropin for milk production in British Friesian dairy cows." *Journal of Agriculture Science* 115:105-112.

West, J. W., B. G. Mullnix, J. C. Johnson, Jr., K. A. Ash, and V. N. Taylor. 1990. "Effects of bovine somatotropin on dry-matter intake, milk yield, body temperature in Holstein and Jersey cows during heat stress." *Journal of Dairy Science* 73:2896-2906.

Wilfond, D. H., K. C. Bachman, H. H. Head, C. J. Wilcox, and G. F. Hartnell. 1989. "Effect of dry-period administration of sometribove upon lactation performance of Holstein cows." *Journal of Dairy Science* 72(Suppl):329. (Abstract).

Wood, R. C., and J. C. Simms. 1991. Personal communication.

Yonkers, R. D., J. W. Richardson, and R. D. Knutson. 1989. "Regional farm level impacts of BST." Working Paper 89-7, Agricultural Food and Policy Center. College Station, TX: Texas A&M University.

Zoa-Mboe, A., H. H. Head, K. C. Bachman, F. Baccari, Jr., and C. J. Wilcox. 1989. "Effect of bovine somatotropin on milk yield and composition, dry-matter intake, and some physiological function of Holstein cows during heat stress." *Journal of Dairy Science* 72:907-916.

5

BST and Animal Health

Dale A. Moore and Lawrence J. Hutchinson

In addition to the efficacy of bovine somatotropin (BST) and its ability to increase milk production in dairy cattle, concerns about animal health as a consequence of its use have been raised. Before any drug or hormone product can be licensed for commercial use in food-producing animals, efficacy and safety to both humans and animals must be demonstrated. It is the purpose of this chapter to describe the studies of BST and dairy cattle which addressed animal health and safety.

Most studies of BST have been designed to assess milk production effects. Several of these production studies have also noted the presence or absence of health effects of dairy cows treated with BST. However, only a few published studies have been specifically designed to examine health effects. Many studies dealing with the health effects of BST treatment are yet to be published. This review will address only those studies already published or presented at scientific meetings.

In reviewing reports of physiologic and health effects of BST, several important characteristics of the experiments influence their applicability to on-farm use of BST:

Dosage level: Was the trial conducted with dosage levels representative of expected use levels?

Duration of administration and time of initiation relative to parturition: Was the timing similar to anticipated usage timing? Did the timing of BST treatment favor or preclude observation of some health effects? E.g., if given late in lactation, effects on fertility would not be as important as if given early in lactation.

Size of groups, trial conditions: Did the trial design permit detection of significant health effects, if present? E.g., were the sample sizes sufficient for detecting significant differences if they did, in fact, exist?

In Appendix Table 5.A we list most of the current trials of BST use that reported health and reproductive events, the time (in days postpartum) of BST treatment, numbers of animals involved, dosages used, if the compound was administered daily or in the sustained-release form, and duration of the study.

Before beginning discussion of the health and reproductive effects of recombinant-BST, it is first necessary to describe the BST compound in general terms. Bovine somatotropin (BST) is a polypeptide (protein) hormone consisting of a chain of 191 amino acids. It normally comes from the anterior pituitary. The major action of this hormone is in the repartitioning of nutrients so as to enhance milk production or growth. Somatotropin is referred to as a "homeorrhetic" hormone. Homeorrhesis is defined as the coordinated changes in the metabolism of body tissues necessary to support a physiologic state (Phipps 1989). Bauman and Currie (1980) proposed that homeorrhetic mechanisms operate on a chronic basis to orchestrate tissue metabolism and involve alterations in tissue response to homeostatic controls. The mechanisms may include:

1. Altered release and/or clearance of a homeostatic signal,
2. Change in blood flow to an organ,
3. Change in sensitivity of a tissue, numbers of receptors, or binding affinity, and
4. Change in tissue responsiveness to a homeostastic signal via a change in the intracellular signal transmitting system.

Somatotropin acts to preserve body protein (Johnsson and Hart 1986), increase the oxidation of fatty acids, inhibit glucose transport in cells, enhance cell division, and promote bone growth (Lean et al. 1990). The most important of these effects are glucose sparing (inhibition of glucose transport into cells) and the promotion of lipolysis or breakdown of fats (diabetogenic action).

Biotechnology, particularly recombinant-DNA technology, has allowed the relatively inexpensive production of highly purified recombinant-BST from *Escherichia coli* cultures. In most preparations, the N-terminal alanine has been replaced with methionine. The growth-promoting and immunological properties of recombinant-BST are identical to those of pituitary-derived BST (Huber 1987). Pituitary-derived BST was found difficult to purify and was usually contaminated with other anterior pituitary hormones, making it less attractive as a milk-yield promotant. Pituitary-derived BST was also in limited supply since it took approximately 200 pituitaries to provide the amount needed for injection of one cow for one day.

The use of BST is predicted to increase milk/cow/year by 6 to 25 percent (McBride et al. 1988). BST apparently does not override physiological mechanisms, but works to increase milk production in cows when management, nutrition, and environment are optimized. When they are not, the response to BST is smaller. In addition, the response to BST decreases as the end of lactation approaches (Phipps 1989).

BST exerts direct effects on some tissues, such as adipose tissue, and also may act through the stimulation of insulin-like growth factors (IGF-I and IGF-II). IGFs are produced in the liver after stimulation by somatotropin and are found to be involved with growth. IGF receptors have also been found in the mammary tissue of cows (Peel and Bauman 1987).

Caveats in Interpreting BST Trials

Because BST trials have differed in important experimental design features such as dose rate, type of preparation (daily vs. sustained-release formulation), length of trial, age of cow, and stage of lactation, the summarization of health effects is difficult. Most studies have had far too few animals to permit statistical analysis for such parameters as days not pregnant postpartum (days postpartum to conception, or days open) and rare health outcomes such as ketosis. Also, many of the studies were reported at professional meetings or published in abstract form without peer review. Furthermore, abstracts do not supply all research methods and must be interpreted with caution. Reports at professional meetings may be of preliminary findings and thus should also be interpreted with caution. A final concern is overall study design and reproducibility of findings. However, with the hundreds of studies of BST use in dairy cows under many different conditions, the safety and efficacy of this product should become apparent.

"The single best indicator of overall health of a lactating dairy cow is milk production. When cows are in poor health or under stress they produce less milk" (Eppard et al. 1989). Administration of BST has resulted in consistently high milk yields across studies which suggests that severe health problems are rare. In this chapter we will address health, reproductive health, and some physiologic effects. A discussion of body condition (weight loss) and nutrition was presented by Muller in Chapter 3.

General Physiologic Effects Associated with BST Use

BST and recombinant-BST alike exert their effects directly on tissues having somatotropin receptors or indirectly through stimulation of production and release of insulin-like growth factors (IGF-I and IGF-II). A number of physiologic effects of BST, such as changes in other hormone concentrations and certain blood constituents, have been studied.

Hormone Changes

Metabolic Hormones. BST apparently coordinates metabolism in favor of milk production. Since milk production is regulated by numerous hormones, investigators have looked at levels of some of these other hormones to determine if BST had an effect on them or if they were involved in the action of BST. Table 5.1 lists the major hormones affecting mammary function. Some of these metabolic hormones assayed by BST investigators were thyroid-stimulating hormone (TSH) and the thyroid hormones (T_3, T_4), insulin, insulin-like growth factors (IGF-I, IGF-II), and prolactin.

The anterior pituitary hormone, TSH, acts on the thyroid gland to stimulate production of thyroxine (T_4) and triiodothyronine (T_3). The general functions of the thyroid hormones are to increase the basal metabolic rate, increase oxygen consumption by the cells, and play a role in the growth, maturation, and function of all cells (Capen and Martin 1989). Surgical removal of the thyroid gland will cause a reduction in milk production (Anderson et al. 1985). Suppression of thyroxine secretion may lead to reduced milk yields while administration of thyroxine stimulates lactational performance. However, the association of BST administration to lactating cattle and changes in thyroid-related hormones is highly variable (Table 5.2). In one study

TABLE 5.1. Major Hormones Affecting Mammary Gland Function.

Endocrine Gland	*Hormone Secreted*	*Function*
Anterior pituitary	Follicle-stimulating hormone	estrogen secretion
	Luteinizing hormone	progesterone secretion
	Prolactin	maintenance of lactation
	Somatotropin	stimulates milk production
	Thyroid-stimulating hormone	stimulates thyroid gland
	Adrenocorticotropic hormone	stimulates adrenal glands to secrete glucocorticoids
Posterior pituitary	Oxytocin	milk ejection
Hypothalamus	Somatotropin-releasing hormone	stimulates somatotropin release
	Somatostatin	inhibits somatotropin release
	Thyroid releasing hormone	stimulates TSH release
	Dopamine	inhibits prolactin release
Thyroid	Thyroxine, triiodothyronine	stimulates oxygen utilization, protein synthesis, milk yield
	Thyrocalcitonin	calcium/phosphorus metabolism
Parathyroid	Parathyroid hormone	calcium/phosphorus metabolism
Pancreas	Insulin	glucose metabolism
Adrenal cortex	Glucocorticoids	maintenance of lactation
	Mineralocorticoids	electrolyte/mineral metabolism
Adrenal medulla	Epinephrine/norepinephrine	inhibition of milk ejection
Ovary	Estradiol	mammary duct growth
	Progesterone	lobule-alveolar growth; inhibition of lactogenesis

Source: Anderson et al. (1985).

on the effect of heat stress (Johnson et al. 1991), the hormone T_3 decreased with BST administration. T_4 was increased in another study (Lanza et al. 1988). In most studies in which thyroid hormones were monitored, there were no changes in thyroid-related hormone levels.

Another metabolic hormone, insulin, is normally at low levels in early lactation and increases almost linearly as lactation proceeds (Anderson et al. 1985). Insulin is involved in stimulating tissue

TABLE 5.2. Effect of Somatotropin on Thyroid-stimulating Hormone (TSH), the Thyroid Hormones T_3 and T_4, and Insulin.

	TSH	*T_3*	*T_4*	*Insulin*
McClean et al. 1988 (Abstract)	NS [a]	NS	NS	-- [b]
Johnson et al. 1991	--	↓↓ [c]	↓↓	NS
Lanza et al. 1988 (Abstract)	NS	↑↑ [d]	NS	--
Capuco et al. 1989	--	NS	NS	--
West et al. 1991	--	NS	NS	--
Soderholm et al. 1988	--	--	--	↑
Staples et al. 1988	--	--	--	↑ over controls
Zoa-Mboe et al. 1989	--	NS	NS	NS
Marcek et al. 1989	NS	--	--	↑ in 1 of 4 cows
Eppard et al. 1987	--	NS	NS	NS
Hutchison et al. 1986 (Abstract)	--	↑↑	↑↑	NS
Sechen et al. 1989	--	--	--	NS
Rowlinson et al. 1989	--	--	NS	NS
Vicini et al. 1990	--	--	NS	--

[a] No significant increases
[b] Not studied
[c] Significant decreases
[d] Significant increases

anabolism (the conversion of nutrients into cell material) and inhibits nutrient mobilization. These functions make insulin regulation a prime target for homeorrhetic regulation (Boyd and Bauman 1989). In normal, nontreated cows, insulin increases as natural growth hormone decreases. BST is observed to alter adipose tissue response to insulin. In the BST trials which examined changes in insulin, insulin was found to be higher in BST-treated animals than in controls in some studies (Vicini et al. 1990, Gallo and Block 1990, Soderholm et al. 1988, Staples and Head 1988, Davis et al. 1987, Schams et al. 1989), while in other studies, blood insulin levels were similar between treatments (Eppard et al. 1987, Hutchison et al. 1986, Sechen et al. 1989, Estrada and Shirley 1990). The reason for the difference could lie with the time of sampling for insulin during the course of the trials. One study (Huber 1987) found that there were minimal changes in the kinetics of insulin after BST administration. Another study found that insulin challenge lowered glucose concentration in the blood to a *lesser extent* in BST-treated cows (Sechen et al. 1989). In addition, the antilipolytic effect of insulin increases during BST treatment (Sechen et al. 1990). Thus, BST effects on insulin secretion have yet to be elucidated, but BST may affect insulin action through receptor function or binding (Boyd and Bauman 1989).

BST may exert its effects through insulin-like growth factor (IGF-I). IGFs have both "acute metabolic and long-term growth promoting effects" (Juskevich and Guyer 1990). IGFs probably act as local tissue growth factors rather than as circulating hormones. Rat studies demonstrated increase in body weight and bone width after infusion with IGF-I. Some scientists found serum IGF-I concentration to be significantly higher in BST-treated animals (Staples and Head 1988, Morbeck et al. 1989, Burton et al. 1989, Davis et al. 1987, Glimm et al. 1988, Prosser et al. 1989, Collier et al. 1990, West et al. 1991, Gallo and Block 1990). In one study, IGF-I was also found in higher concentrations in the milk of BST-treated cows compared to controls (Prosser et al. 1989). However, IGF-I levels were not higher than physiological levels found normally in human milk. BST administration was reported to alter the distribution of stainable IGF-I material in mammary tissue and may therefore indirectly result in the growth or function of mammary epithelial cells (Glimm et al. 1988). In fact, serum IGF-I concentration increased linearly with increasing dosage of BST (Morbeck et al. 1989). In a study reported by de Boer and

Kennelly (1988) the correlation between BST and IGF-I was negative, but that between BST and milk yield was positive.

Prolactin concentrations in the blood are positively correlated with milk yield. However, in ruminants, suppression of prolactin secretion does not affect milk production (Anderson et al. 1985). Differences in prolactin concentrations following BST administration were not found in four studies (Lanza et al. 1988, Zoa-Mboe et al. 1989, Eppard et al. 1987, Johnson et al. 1991) while significantly increased concentrations were found in studies by Hutchison et al. (1986), Bines et al. (1980), and Peel et al. (1981).

Normally, at the onset of lactation, there is an increased lipolytic response to epinephrine by the adipose tissue (Anderson et al. 1985). An investigation of the effect of BST administration on the ability of epinephrine to stimulate lipolysis in lactating cattle determined that release of nonesterified fatty acids into the blood of BST-treated cows was greater than that of control cows (Sechen et al. 1990).

Stress-related Hormones. Administration of therapeutic doses of adrenocorticotropic hormone (ACTH), an anterior pituitary hormone, or of adrenal glucocorticoids such as cortisol, almost always leads to reduced milk secretion. Basal secretion of glucocorticoids does not normally change during lactation and is actually necessary for the maintenance and initiation of lactation (Anderson et al. 1985). ACTH and cortisol, normally associated with stress, were found in the same concentrations in BST-treated cows and control cows as well as pre- and post-treatment with BST in the same cows (Jenny et al. 1988, Lanza et al. 1988, West et al. 1989, Marcek et al. 1989, West et al. 1991). Thus, BST administration apparently does not cause a change in serum concentration of stress-related hormones.

Reproductive Hormones. The major reproductive hormones are estrogens and progesterone, produced in the ovaries, and the pituitary gonadotropins, luteinizing hormone (LH), and follicle-stimulating hormone (FSH). Estrogens have some inhibitory effects on milk secretion, although low dosages may actually increase milk secretion. Progesterone apparently does not affect milk secretion (Anderson et al. 1985). Curves showing normal changes in reproductive hormone concentrations during lactation are presented in Figure 5.1. Reproductive hormones have been measured in a few BST studies. Luteinizing hormone (LH), an anterior pituitary hormone, is involved with the stimulation of progesterone secretion from the corpus luteum of the ovary. LH release is stimulated by gonadotropin-releasing hormone (GnRH). LH does not directly affect mammary development or function. Moseley (1989) found that the GnRH-induced LH response was not affected by elevated serum BST. In another study, the LH pulse frequency was found to be more rapid in BST-treated cows, but the amplitude of LH pulses was not different (Schemm et al. 1990).

Receptors for BST have been isolated from the bovine ovary (Tanner and Hauser 1989). Granulosa cells in the bovine ovary apparently produce IGF-I that acts synergistically with FSH to increase the production of estrogen and progesterone (Spicer et al. 1990). Schemm et al. (1990) reported that the average progesterone concentration was higher in BST-treated cows, but that BST had no influence on estradiol (estrogen) concentrations, and the length of the cows' reproductive cycles were unchanged. Gallo and Block (1990) also found a higher peak progesterone output during the first two estrous cycles and during pregnancy in BST-treated cows. This may be due to an ovarian IGF-I-mediated mechanism that was demonstrated by Spicer et al. (1990). Normal changes in milk-progesterone concentration in another study were interpreted to indicate normal estrous cycling prior to and after BST treatment (Eppard et al. 1987).

Minerals and Electrolytes

BST treatment has been reported to affect mineral metabolism. BST treatment has increased gut absorption of calcium, increased accumulation of calcium into bone (bone growth) and also disappearance or resorption of calcium from bone (Boyd and Bauman 1989). Of five BST trials that examined the blood levels of calcium and phosphorus (Eppard et al. 1987, Soderholm et al. 1988, Lanza et al. 1988, West et al. 1988, McGuffey et al. 1989), only that of Lanza et al. (1988) reported a higher calcium concentration in BST-treated animals compared to controls. Phosphorus levels were higher in BST-treated cows studied by Whitaker et al. (1989); calcium and magnesium levels in these cows were normal. Magnesium was found to be lower in BST-treated cows than in controls studied by Lanza et al. (1988). Thus, the results reported for mineral balance after BST treatment vary and need to be interpreted with caution.

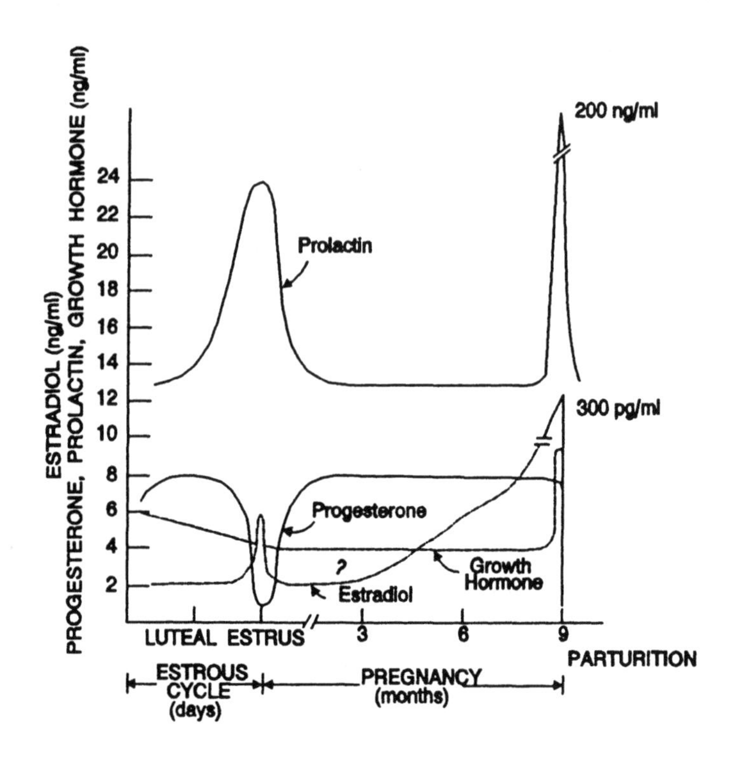

FIGURE 5.1. Serum concentrations of estradiol, progesterone, prolactin, and growth hormone in cattle during the estrous cycle and pregnancy (Larson et al. 1985, with permission).

Sodium, chloride, and potassium were not significantly different between BST-treated and control cows (McGuffey et al. 1989, Soderholm et al. 1988). Other workers reported significant differences in electrolyte concentrations, but gave no values.

Body Temperature

The increase in milk yield after BST administration is not as great when cows are heat-stressed as when cows are in a normal thermal environment (Zoa-Mboe et al. 1989). Because BST regulates metabolism and increases dry-matter intake, some investigators have measured changes in body temperature during administration of BST. All body temperatures in several studies were within the normal range

(Soderholm et al. 1988, Eppard et al. 1987, Johnson et al. 1991, Weller et al. 1990) In five studies which attempted to evaluate the effects of heat-stress on BST-treated cattle, cows in three of the five studies showed significant increases in rectal temperature over controls, especially if the cows did not have access to shade (Staples and Head 1988, Zoa-Mboe et al. 1989, West et al. 1990). The increase in body temperature of Jerseys administered BST was greater than that of Holsteins (West et al. 1990). The other studies demonstrated no significant increases in rectal temperature with increasing ambient temperatures (Johnson et al. 1991, Mohammed and Johnson 1985). BST was less effective in increasing milk production at high ambient temperatures with high humidity in one study (West et al. 1990) but still resulted in milk yields 20 to 35 percent higher than controls.

A toxicity study using 30 grams of the sustained-release formulation of BST resulted in significantly higher mean rectal temperatures of treated animals compared to controls (Vicini et al. 1990).

Only one research team evaluated cold stress and BST effectiveness in increasing milk yields. Li et al. (1988) found that the BST stimulation of milk production is maintained even with cold ambient temperatures. Thus, cows did not have to divert energy to maintenance of body temperature at the expense of milk production during BST administration.

Effects of BST on Body Systems

Although all the body's physiologic functions work in a coordinated fashion, we can compartmentalize our view of the body by dividing it into organ systems. An organ system performs a specific function for the maintenance of the animal. Our use of the term "body system" is rather loose in this section, but it does provide us a framework to discuss the BST literature with regard to certain changes in function.

In most BST trials it is difficult to separate direct effects of BST on a body system from effects associated with increased milk production. The following is a brief review of what has been published with regard to BST administration and its association with changes in the cardiovascular and hemolymphatic (blood) systems, metabolic diseases, mammary-gland health, other infectious diseases, and immune and musculoskeletal systems. Because reproductive health has been examined more extensively than other health problems, this subject is covered in another section.

Cardiovascular/Pulmonary Changes

Increased milk yields should reflect increased flow of blood for carrying milk-precursors to the mammary gland. Control of mammary blood flow may be one way to control nutrient partitioning. Cardiac output was reported to be 10 percent higher and mammary blood flow increased by 35 percent in BST-treated cows studied by Davis et al. (1983). Higher cardiac output was also reported in a study by Tanner and Hauser (1989). Changes in cardiac output apparently occur soon after BST administration, before any resulting increase in milk yield Johnsson and Hart (1986). Increased mammary plasma flow was also reported by Fullerton et al. (1989). How blood flow to the mammary gland is regulated is not well-understood, but it is generally accepted that carbon dioxide concentration is the major regulator of blood flow through a tissue.

Heart rates were monitored regularly in several studies. Heart rate was slighter higher in BST-treated cows at high dose rates but was still within normal range (Soderholm et al. 1988, Eppard et al. 1987). Heart rates were unaffected in one study (Weller et al. 1990).

Pulmonary function was monitored only in a few studies. West et al. (1990) noted a significant decrease in carbon dioxide concentration in the blood of BST-treated cows during heat-stress compared to control heat-stressed cows. This may have been due to an increased respiratory rate, because BST-treated cows had higher mean rectal temperatures compared to control cows. Normally, respiratory rate will increase in higher ambient temperatures, reducing blood carbon dioxide concentration. Respiratory rate was also found to be higher in BST-treated cows under heat-stress in a study by Staples and Head (1988). Respiratory rate was variable in one study (Soderholm et al. 1988) and was not different from that of control animals in another (Weller et al. 1990).

Hemolymphatic (Blood) System

Many studies of BST and milk yield included analyses of blood samples for complete blood counts (CBC). Lower hematocrit or packed red blood cell volume (PCV) values were fairly consistent in studies of BST-treated cows reported by Vicini et al. (1990), Zoa-Mboe et al. (1989), Eppard et al. (1987), Soderholm et al. (1988), Marcek et al. (1989), and Fullerton et al. (1989). Lower PCV values may indicate an anemia or an elevated ratio of plasma to red blood cells. Marcek et al. (1989) reported a lower hemoglobin value in their study of BST-treated cows. McGuffy et al. (1989) and Gallo and Block (1990) found no changes in PCV values. Hemoglobin concentration is directly related to PCV except in diseases in which utilization of iron is defective. Interpretation of the PCV must always be made in light of the water balance of the animal (Schalm et al. 1975). The most likely explanation for a lower PCV in BST-treated cows is through hemodilution, or an increased plasma-fluid volume compared to blood cell constituents. There is normally a difference in size of the blood volume between dry and lactating animals. For example, initiation of lactation leads to an increase in the size of the vascular bed, followed by an increase in blood volume (Schalm et al. 1975). PCV has been reported to be significantly lower during early lactation and increases through the dry period (Kradel et al. 1975, Esievo and Moore 1979, Rowlands et al. 1979).

Metabolic Diseases

Since BST is believed to act by changing the metabolism of cows and repartitioning nutrients towards production of milk, investigators have attempted to detect changes in incidence of metabolic disease in BST-treated cows compared to controls. Burton et al. (1990) merely stated "No effect on metabolic diseases" but most reports provided information on the specific diseases monitored.

There has been concern that BST use would result in higher rates of ketosis. Most of this concern has come from one study of pituitary-derived growth hormone which resulted in high levels of ketones in the blood of two BST-treated cows, however there was no clinical ketosis (Kronfeld 1965). Clinical ketosis, milk fever, and fatty liver are fairly rare events and a study would require a large number of animals to detect a difference if one existed. Most studies have not involved the number of cows usually required to detect true differences in ketosis, milk fever, or fatty liver. However, the lack of reports of ketosis in the very large number of cows (>10,000) in all studies that have received BST suggests that BST administration does not cause ketosis in well-managed dairy cows.

Ketosis and milk fever were not seen in the small study reported by McBride et al. (1988), in a New York study (Eppard et al. 1987), nor in a few other studies in which health events were monitored (Bauman et al. 1989, Lormore et al. 1990, and Weller et al. 1990). Serum betahydroxybutyrate increases in cows with subclinical ketosis. Several studies have demonstrated that this indicator of subclinical ketosis is not increased after BST administration (Eppard et al. 1987, Lanza et al. 1988, Dhiman et al. 1988, Hutchison et al. 1986, Vicini et al. 1990). BST also did not change the *in vitro* production of ketones or carbon dioxide from butyrate or palmitate (Pocius and Herbein 1986). Increased ß-hydroxybutyrate in BST-treated cows, compared to controls, was found in two studies (Whitaker et al. 1989, Rowlinson et al. 1989).

Liver enzymes are released into the blood after damage to the liver cells and liver damage is often assessed by the levels of these enzymes in the blood. In the bovine, these enzymes include sorbital dehydrogenase (SDH), alkaline phosphatase, and serumglutamic oxaloacetic transaminase (SGOT). Liver enzyme levels of BST-treated and control cows were reported to be within normal limits in three studies (Lanza et al. 1988, McGuffey et al. 1989, Gallo and Block 1990). In another study, alkaline phosphatase levels were increased in one out of four cows (Marcek et al. 1989). Increased serum bilirubin can sometimes indicate a problem in liver function. Serum bilirubin was found to be slightly increased in BST-treated cows given very high dosages (Vicini et al. 1988).

Blood urea nitrogen (BUN), if low in the ruminant, can indicate low protein in the diet or protein metabolism problems. If high, it may indicate kidney malfunction or consumption of relatively large amounts of readily degradable protein. A lower BUN might also indicate a more efficient protein utilization. BUN was not affected in BST-treated cows in three studies (Soderholm et al. 1988, Dhiman et al. 1988, Estrada and Shirley 1990). In one study, BUN was reported to decrease early in the BST treatment period and returned to control levels during the remainder of the lactation (Lanza et al. 1988). BUN decreased following BST treatments in a number of studies (Marcek et al. 1989, Whitaker et al. 1989, McGuffey et al. 1989, Vicini et al. 1988, West et al. 1991) and BUN concentrations were significantly depressed on days 8 and 14 of a 14-day toxicity study (Vicini et al. 1990).

Mammary Gland

Mastitis is generally considered to be the most economically important disease of dairy cows. Milk somatic cell counts (SCC) provide an indirect measure of mastitis in cattle and influence milk quality and cheese yield. Many BST studies have monitored SCC levels and/or clinical signs of mastitis. At least one study has examined the response of BST-treated cows to experimental mastitis challenge.

Lissemore et al. (1991) reported that the incidence of clinical mastitis was higher in BST-treated cows compared to controls, but that there was no difference in geometric mean SCCs. Incidence of clinical mastitis was not higher in BST-treated cows in most studies (Eppard et al. 1987, Rowe-Bechtel et al. 1988, Palmquist 1988, Lamb et al. 1988, Hard et al. 1988, Chalupa et al. 1988, Thomas et al. 1989, Weller et al. 1990, Burton et al. 1990, Whitaker et al. 1988, Thomas et al. 1991). SCC of the individual cow is widely used as a measure of subclinical mastitis. Average milk SCCs were higher among BST-treated cows compared to controls in two studies (Weller et al. 1990, Rijpkema et al. 1987). However, average SCC did not differ in most trials (Eppard et al. 1987, Jenny et al. 1988, Wilfond et al. 1989, Bunn et al. 1988, Elvinger et al. 1987, Hemken et al. 1987, Rowe-Bechtel et al. 1988, Munneke et al. 1988, Hard et al. 1988, Eppard et al. 1987, Cole et al. 1988, Aguilar et al. 1988, Rock et al. 1989, Thomas et al. 1989, Galton 1989, Franson et al. 1989, McGuffey et al. 1989, Thomas et al. 1991). In a review of sustained-release BST trials by Peel and Bauman (1987), treated cows had higher SCC but these differences were significant in only two of eight studies. Phipps (1989) summarized the mastitis and SCC results of BST investigations conducted in a number of different countries. SCCs tended to be higher in treated cows compared to controls, although there were no increases in the incidence of clinical mastitis (Phipps 1989).

Short-term treatments with BST might be beneficial in treating *E. coli* mastitis in postparturient cows and goats (Burvenich et al. 1988, Burvenich et al. 1989). Investigators found that BST treatment did not restore milk yield to its original level, but did increase milk yield above that usually expected following recovery from a coliform mastitis infection. They thought that BST somehow may influence the restoration of the blood/milk barrier after clinical coliform mastitis. It has also been observed that alpha-lactalbumin, an indicator of milk synthesis, was significantly increased in the milk of treated cows, reflecting a stimulation of milk synthesis in the mastitic gland (Roets et al. 1988).

Infectious Diseases and Immunology

If BST puts undue stresses on cows, one might assume that cows treated with BST would have a higher incidence of infectious diseases or depressed immune responses. Thus, some investigators have monitored infectious diseases in their studies and a few have looked at immune-system effects.

One investigator monitored infectious diseases and reported no effect on the incidence of diseases (Burton et al. 1990). There are no reports of higher incidence of diseases such as pneumonia among BST-treated cattle compared to controls.

A few investigators have looked at the response of defense-system cells after BST treatment. One study (Heyneman et al. 1989) reported that "Elevated blood levels of somatotropin may favor the proliferation and generation of a neutrophil population which is 'primed' for an augmented oxidative response." This primed oxidative response would enhance killing of bacteria by the neutrophils. Burton et al. (1991) suggested that BST may augment cow immunity. They demonstrated that the lymphocytes from the blood of treated cows responded to stimulation by mitogen with a higher proliferative responsiveness than cells from control cows. This proliferative response to mitogen has also been demonstrated in rats (Davila et al. 1987).

In one study of BST-treated cows, no significant effect on the immune system was detected, as monitored by differences in proliferative responses induced by phytohemagglutinin (PHA), interleukin 2 (IL-2) production, and phagocytosis by white blood cells (polymorphonuclear cells) (Estrada et al. 1989). However, in another study, BST administration was thought to elevate the capacity of T-lymphocytes to proliferate upon antigenic challenge (Burton et al. 1989). These immuno-enhancing properties are apparently independent of BST-induced elevations in circulating IGF-I (Burton et al. 1989). Another investigator (Elvinger et al. 1990) suggested that BST administered *in vivo* enhances resistance of lymphocytes to *in vitro* heat-stress. Thus, rather than an inhibition of the animal's immune system, there is some evidence for BST administration's enhancement of some parts of its immune system. BST had no apparent detrimental effect on humoral immunity as measured by blood antibody concentrations (Burton et al. 1991).

Musculoskeletal System--Lameness, Injection-site Reaction

No significant difference in the incidence of lameness or hoof injuries was found in any study that monitored these ailments (Eppard et al. 1987, Phipps et al. 1988, Chalupa et al. 1988, Burton et al. 1990, Lormore et al. 1990). Normal bone growth, development, density, and periosteal activity were noted by one investigator (Phipps 1989) indicating that bones did not become depleted of calcium or phosphorus during treatment with BST.

Phipps et al. (1988) reported a mild injection-site reaction with the sustained-release BST that resolved within 2 to 4 days. Abscesses or other complications were not noted. Irritant swelling at injection sites were noted by two other investigators. One used a very high-dose daily injection (Marcek et al. 1989) and the other used the sustained-release formulation (Whitaker et al. 1988).

BST-associated Changes in Reproduction

In a number of studies, initiation of BST administration began at approximately 100 days postpartum or was delayed until cows were diagnosed pregnant. In these trials, no opportunity was available to examine effects on estrous cycles or fertility since most cows should become pregnant by 100 days postpartum. In other trials, BST administration was initiated in early lactation, and estrous cycle and/or fertility effects could be observed. In some of the latter studies, production-associated weight loss may have impaired reproductive performance. Although most trials with BST did not include the number of cows required to determine significant differences in normal reproductive indices, no effects or variable effects on fertility were noted in most trials in which fertility was monitored. Phipps et al. (1988) has reported results suggesting that the effects of BST on reproduction are correlated more with milk yield than with the administration of BST. One study stated that no differences in reproduction indices of BST-treated cows and of controls were detected, but specifics were not provided (Mattos et al. 1989). For a list of some of the studies that reported reproductive outcomes of BST-treated cows and the reproductive indices they measured, see Table 5.3.

For those studies in which reproductive performance was a consideration, treatment with BST began sometime after the first estrus and before the second estrus. This is the time during lactation when cows are first being observed for estrus. When compared to controls, conception rates have been reported to be lower in BST-treated cows in some studies (Bunn et al. 1988, McBride et al. 1988, Pell et al. 1988, Bauman et al. 1989, Chalupa et al. 1987, Weller et al. 1990, Lormore et al. 1990, Huber et al. 1988, Samuels et al. 1988, Thomas et al. 1989); while in other studies, the conception rates did not differ or were variable among treatment groups (Rajamahendran et al. 1989, deBoer and Kennelly 1989, Hansen et al. 1989, McDaniel et al. 1989, Cleale et al. 1989, Eppard et al. 1987, Chalupa et al. 1989, Thomas et al. 1987a, Phipps et al. 1988, Grings et al. 1990, Rowlinson et al. 1989). One investigator reported normal conception rates after the first three services, but reported fewer established pregnancies, or higher embryo loss as detected by progesterone analysis after high BST dosages (Thomas et al. 1987a, Thomas et al. 1987b). No early embryonic deaths were observed by Eppard et al. (1987). Percent pregnant at time of drying off (cessation of milking) was not significantly different among treatment groups (Cole et al. 1989). Because of the binomial variation in the estimation of conception rates (a cow *was* or *was not* pregnant) large numbers of observations are required to detect a statistically significant difference.

Differences in average days open (days postpartum until conception) between BST-treated cows and controls were also somewhat variable, probably reflecting the inadequate sample sizes in the studies. No difference in average days open was reported for about one-half the studies that monitored average days open (Eppard et al. 1987, deBoer and Kennelly 1989, Rajamahendran et al. 1989, Murphy et al. 1989,

TABLE 5.3. Differences in Some Reproductive Outcomes Between BST-treated and Control Cows.

Investigators	*CR* [a]	*Days Open*	*S/C* [b]	*FSCR* [c]	*DTFS* [d]
Eppard et al. 1987	NS [e]	↓ [f]	↓	NS	-- [g]
Bauman et al. 1989	NS	--	--	--	--
Thomas et al. 1987b	NS	--	--	--	
Burton et al. 1987	--	↑ [h]	--	--	NS
Chalupa et al. 1987	↓	--	NS	--	--
Peel et al. 1981	↓	↑	↑	--	--
Huber et al. 1988	↓	↑	--	--	--
Samuels et al. 1988	↓	--	--	--	--
Bunn et al. 1988	↓	--	--	--	--
Chalupa et al. 1989	NS	NS	--	--	--
Wilfond et al. 1989	--	NS	--	--	--
Cleale et al. 1989	NS	↑	↑	--	--
McDaniel et al. 1989	NS	--	--	--	--
Hansen et al. 1989	NS	↑	↑	--	--
deBoer&Kennelly 1989	NS	NS	NS	--	--
Rajamahendran et al. 1989	NS	NS	--	--	--
Murphy et al. 1989	--	NS	NS	--	NS
Thomas et al. 1989	↓	--	--	--	--
Weller et al. 1990	↓	↑	↑	--	--
White et al. 1988	--	~↑	~↑	--	--
Morbeck et al 1989	--	NS	--	--	--
Burton et al. 1990	--	NS	--	--	--
Furniss et al. 1988	--	NS	--	--	--
Thomas et al. 1987a	NS	--	--	--	--
Whitaker et al. 1988	--	--	↑	--	↑
Phipps et al 1988	NS	↑	--	--	--
Lormore et al 1990	↓	↑	↑	↑	~↑

[a] Conception rate
[b] Services per conception
[c] First-service conception rate
[d] Days to first service
[e] No significant difference
[f] Decrease
[g] Not monitored
[h] Increase

White et al. 1988, Wilfond et al. 1989, Morbeck et al. 1989, Moseley 1989, Chalupa et al. 1989, Samuels et al. 1988, Furniss et al. 1988, Grings et al. 1990). Longer average days-open periods in BST-treated cows compared to controls were reported in the other half of the studies (Hansen et al. 1989, Cleale et al. 1989, Huber et al. 1988, Thomas et al. 1987a, Burton et al. 1987, White et al. 1988, Pell et al. 1988, Weller et al. 1990, Phipps et al. 1988, Lormore et al. 1990, Elvinger et al. 1987). In one study, where cows were pooled by production rather than treatment, production rate had a greater influence on days open than did BST administration (Weller et al. 1990). However, a dose-response was noted in three other studies: at higher BST-dose rates, the days-open period increased (Thomas et al. 1987a, Burton et al. 1987, Chalupa et al. 1988). When started early postpartum, BST administration was more likely to influence days open compared to when started later in lactation (e.g. 84 days) (Eppard et al. 1987). In one study (Cole et al. 1989), days-open was highest for primiparous cows receiving 750 mg BST in sustained-release formulation every 14 days compared to cows receiving 0, 250 or 500 mg every 14 days throughout lactation.

Calving interval, related to days open, was reported in several studies. Calving interval in some studies was longer in BST-treated cows (Huber et al. 1988, White et al. 1988, Weller et al. 1990) but not significantly different in three other studies (White et al. 1988, Butterwick et al. 1988, Rowlinson et al. 1989). Calving interval was extended 27 to 68 days compared to controls (McGuffey et al. 1991).

Services per conception was reported to be unchanged in some studies (Chalupa et al. 1987, deBoer and Kennelly 1989, Murphy et al. 1989, Cole et al. 1989, White et al. 1988, Moseley 1989, Morbeck et al. 1989) while higher in other studies (Pell et al. 1988, White et al. 1988, Whitaker et al. 1988, Weller et al. 1990, Cleale et al. 1989, Lormore et al. 1990, Hansen et al. 1989). Total number of inseminations for BST-treated cows was twice that of controls (McGuffey et al. 1991). First-service conception rate was monitored in two studies and was found to be lower for BST-treated cows (Lormore et al. 1990) in one and no different in the other (Eppard et al. 1987). Days to first service were increased in four of five studies that monitored it (Moseley 1989, Burton et al. 1990, Whitaker et al. 1988, Lormore et al. 1990) and was not increased in one (Murphy et al. 1989).

Several investigators have looked at changes in the estrous cycle in BST-treated and control cows. Milk-progesterone analysis demonstrated normal cycling in BST-treated cows in one study (Eppard et al. 1987). Two studies concluded that BST treatment may have suppressed expression of estrus at 20 mg/d dosages of BST (Morbeck et al. 1989, Burton et al. 1990). At a BST dose of 20.6 mg/d, investigators found that an increased number of therapeutic injections of gonadotropin-releasing hormone for ovarian cysts and prostaglandins for the induction of estrus were required to attain normal estrous cycles (Burton et al. 1990). The investigators wondered if this was related to energy balance at the time of the first and second estrous cycles postpartum. In a different study, which monitored energy balance and abnormal estrous cycles in cows, it was observed that anestrus cows and cows with late initiation of luteal activity (between day 40 and 60 compared to day 40) used more energy from body reserves for milk production during the first 2 weeks of lactation (Staples et al. 1990) than did cows with normal initiation of luteal activity. This demonstrated that the degree to which cows divert stored energy to milk production could influence their reproductive cycles. Another study using three-times-per-week milk-progesterone analysis found no differences between BST-treated and control cows in mean cycle lengths, mean peak luteal activity, or number of days to first ovulation.

In another study, estrus detection rate after the first ovulation decreased linearly with increasing doses of BST (Morbeck et al. 1989). It was thought that BST treatment may have suppressed expression of estrus. Days between detected estruses increased linearly. In another study, there was a longer time between parturition and established pregnancy (Thomas et al. 1987a). This could have been caused by early embryonic deaths as well as by changes in estrous cycles.

"Time of introduction, dose rate and plane of nutrition, all of which would affect energy status, are likely to influence reproductive performance of treated cows" (Phipps 1989). It was demonstrated that decreased reproductive efficiency is associated with increased milk energy output. This is the most likely mechanism whereby BST affects reproduction. Energy deficit has an important negative influence on the reestablishment of normal LH pulse patterns necessary for initiating ovarian follicular development. Studies that have examined endocrine (hormone) profiles and follicle sizes have not detected differences due to BST. The differences appear when reproduction indices such as conception rate, average days open, services per conception, etc., are examined.

Ferguson and Skidmore (1989) have summarized some of the litera-ture on BST effects on reproduction and health using data reported in papers and abstracts from 1985-1989. They used a statistical model to control for BST preparation, dose, injection frequency, time postpartum at initiation of BST treatment, and dosage frequency. From DHIA data, higher-producing herds had lower conception rate (CR) than lower producing herds (62 percent vs. 52 percent) and higher producing cows had lower CR than lower-producing cows (<13,000 lbs. of milk, CR = 48.5%; >17,000 lbs. of milk, CR = 38.6%). They concluded that days open increased at higher doses of BST. They projected that BST use will lower pregnancy rates under current reproductive-management schemes. It is likely that management, particularly nutri-tion, will need to change with the adoption of BST use on dairies. Breeding may have to be delayed in cows receiving BST until after 150 days postpartum, when BST has less influence on milk yield and nutrient partitioning than it does early in lactation (Ferguson and Skidmore 1989). Initiation of BST use at 100 days postpartum or after the cow is diagnosed pregnant would avoid possible negative effects of BST on reproductive performance.

Other Effects of BST in Dairy Cattle

Calving and Offspring

Ease of calving, incidence of parturient paresis, and incidence of retained placentas did not vary among treatment groups (Cole et al. 1988). No increase in dystocia was found by Wilfond et al. (1989). One study looked at the incidence of abortions and found no difference between BST-treated and control cows (McDaniel et al. 1989).

A number of studies examined the calves born to BST-treated cows. No effect on length of gestation was found (Eppard et al. 1987, Anderson et al. 1989, Rowlinson et al. 1989) and no effect on the birthweight of the calves was reported (Eppard et al. 1987, Thomas et al. 1987a, Hansen et al. 1989, Weller et al. 1990, Thomas et al. 1987b). One surprising finding, however, was a high incidence of twinning in some studies (Hansen et al. 1989, Weller et al. 1990, Butterwick et al. 1988, McGuffey et al. 1991). This was found only in primiparous animals in one study (Butterwick et al. 1988). One investigator noted that single calves of treated cows tended to weigh less at birth but gained more in the first 9 weeks than calves of control cows (Anderson et al. 1989). At 9 weeks of age, weight differences were not found between these two groups of calves. No effect on the growth of calves was reported in studies by Eppard et al. (1987) or Marcek et al. (1989).

Subsequent Lactation

In studies which looked at second lactations after BST treatment, there was no difference in response to BST during the second lactation compared with the first lactation. These workers also found no adverse effects on reproduction or SCC in second lactations (Hemken et al. 1987, Chalupa et al. 1989, Weller et al. 1990). However, cows in their second lactation with BST administration initiated at week 2 required significantly more services per conception and had longer calving intervals. This was likely due to their lower body condition (Rowlinson et al. 1989).

Culling

In a study of survival rates and culling of BST-treated cattle, reasons for culling were different but there was no evidence of lower rates of survival of BST-treated cows. A greater percentage of treated cows were culled for reproduction compared to controls (McDaniel et al. 1990). In another study, the reasons for culling were related to reproduction or mastitis (Burton et al. 1990).

Conclusions

A number of BST trials examined the effects of this recombinantly-derived hormone on levels of milk production. In some studies, the presence or absence of health effects have also been noted. Few trials were designed specifically to measure possible effects of BST on cattle health. A wide variety of physiologic parameters have been evaluated for possible effects of BST. Increases in circulating levels of triiodothyronine (T_3) and insulin have been noted. Most studies detected no differences in circulating levels of adrenocortical, reproductive, or other hormones. The response of BST-treated cows to high environmental temperatures is similar to that of untreated cows, but milk enhancement with BST is less at higher temperatures.

Changes in several body systems have been noted with BST use. Cardiac output and mammary blood flow is increased. Hematocrit, or packed cell volume, is decreased, as is blood urea nitrogen. Immune reactions of BST-treated cows have been evaluated both by examination of immune mechanisms and by clinical response to natural or experimental infections. The oxidative response of neutrophils, a measure of the white blood cells' ability to fight infection, is enhanced in BST-treated cows.

Most studies have reported no difference in clinical mastitis or somatic cell count between BST-treated and control cows. Limited information indicates that the immune response to or the ability of the animal to recover from *E. coli* mastitis is enhanced by BST treatment.

Possible reproductive effects of BST administration have been of considerable interest and concern to research workers. In a number of studies, BST treatment was initiated about 100 days postpartum or after cows were confirmed pregnant, so no opportunity for observing fertility effects was afforded. In those trials in which BST administration was initiated in early lactation, the trials showing no effect and showing effects on fertility were about equal. When effects on fertility were noted, these were generally associated with increased milk production, negative energy balance, and weight loss during early lactation. Some workers noted an increased incidence of twinning in BST-treated cows.

When animal-health effects have been documented in BST studies, they have generally been shown to be secondary to increased milk production, indicating the importance of excellent nutrition and management if BST is used to enhance production. The results of other animal studies should be reviewed upon their publication. Many of these studies are currently being reviewed by the Food and Drug Administration, the agency responsible for the approval of BST for commercial use in milk production.

Acknowledgments

The authors gratefully acknowledge the review of this manuscript by Dr. Brian Crooker, University of Minnesota; Dr. Michael O'Connor,

The Pennsylvania State University; and Dr. James Ferguson, University of Pennsylvania.

References

Aguilar, A. A., D. C. Jordan, J. D. Olson, C. Bailey, and G. Hartnell. 1988. "A short-term study evaluating the galactopoietic effects of the administration of sometribove (recombinant methionyl bovine somatotropin) in high producing dairy cows milked three times per day." *Journal of Dairy Science* 71(Suppl):208. (Abstract)

Anderson, M. J., R. C. Lamb, R. J. Callan, G. F. Hartnell, R. G. Hoffman, L. Kung, Jr., and S. E. Franson. 1989. "Effect of sometribove (recombinant methionyl bovine somatotropin) on gestation length and on body measurements, growth, and blood chemistries of calves whose dams were treated with sometribove." *Journal of Dairy Science* 72(Suppl):327. (Abstract)

Anderson, R. R., R. J. Collier, A. J. Guidry, C. W. Heald, R. Jenness, B. L. Larson, and H. A. Tucker. 1985. *Lactation*. Ames, IA: Iowa State University Press.

Arambel, M. J., S. O. Oellerman, R. C. Lamb, and G. A. Green. 1989. "The effect of sometribove (recombinant methionyl bovine somatotropin) on production response in lactating dairy cows: A field trial." *Journal of Dairy Science* 72(Suppl):451. (Abstract)

Bauman, D. E., and W. B. Currie. 1980. "Partitioning of nutrients during pregnancy and lactation: A review of mechanisms involving homeostasis and homeorrhesis." *Journal of Dairy Science* 63:1524-1529.

Bauman, D. E., D. L. Hard, B. A. Crooker, M. S. Partridge, K. Garrick, L. D. Sandles, H. N. Erb, S. E. Frabson, G. F. Hartnell, and R. L. Hintz. 1989. "Long-term evaluation of a prolonged-release formulation of N-methionyl bovine somatotropin in lactating dairy cows." *Journal of Dairy Science* 72:642-651.

Bines, J. A., I. C. Hart, and S. V. Morant. 1980. "Endocrine control of energy metabolism in the cow: the effect on milk yield and levels of some blood constituents of injecting growth hormone and growth hormone fragments." *British Journal of Nutrition* 43:179-188.

Boyd, R. D., and D. E. Bauman. 1989. "Mechanisms of action for somatotropin in growth." Pp 257-293 in D. R. Campion, G. J. Hausman, and R. J. Martin (eds). *Animal Growth Regulation*. New York: Plenum Press.

Bunn, K. B., B. F. Jenny, F. E. Pardue, G. S. Bryant, and D. W. Rock. 1988. "Effect of sustained-release bovine somatotropin (BST) on reproduction and mammary health of dairy cows." *Journal of Dairy Science* 71(Suppl):325-326. (Abstract)

Burton, J. H., B. W. McBride, K. Batean, G. K. Macleod, and R. G Eggert. 1987. "Recombinant bovine somatotropin: effects on production and reproduction in lactating cows." *Journal of Dairy Science* 70(Suppl):175. (Abstract)

Burton, J. L., B. W. McBride, J. H. Burton, and R. G. Eggert. 1990. "Health and reproductive performance of dairy cows treated for up to two consecutive lactations with bovine somatotropin." *Journal of Dairy Science* 73:3258-3265.

Burton, J. L., B. W. McBride, B. W. Kennedy, and J. H. Burton. 1989a. "Exogenous rBST enhances bovine T-lymphocyte responsiveness to mitogen." *Journal of Dairy Science* 72(Suppl):150. (Abstract)

Burton, J. L., B. W. McBride, B. W. Kennedy, T. H. Elsasser, and J. H. Burton. 1989b. "Exogenous rbST, circulating levels of IGF-I and Bovine T-lyphocyte function." *Journal of Dairy Science* 72(Suppl): 150. (Abstract)

Burton, J. L., B. W. McBride, B. W. Kennedy, J. H. Burton, T. H. Elsasser, and B. Woodward. 1991. "Influence of exogenous bovine somatotropin on the responsiveness of peripheral blood lymphocytes to mitogen." *Journal of Dairy Science* 74:916-928.

Burton, J. L., B. W. McBride, B. W. Kennedy, J. H. Burton, T. H. Elsasser, and B. Woodward. 1991. "Serum immunoglobin profiles of dairy cows chronically treated with recombinant bovine somatotropin." *Journal of Dairy Science* 74:1589-1598.

Burvenich, C., A. M. Massart-Leen, G. Vandeputte-Van Messom, E. Roets, and G. Kiss. 1989. "Effect of recombinant bovine somatotropin in endotoxin induced mastitis in lactating dairy goats." *Archives International Physiology and Biochemistry* 97:91-95.

Burvenich, C., G. Vandeputte-Van Messom, E. Roets, J. Fabry, and A. M. Massart-Leen. 1988. "Effect of bovine somatotropin on milk yield and milk composition in periparturient cows experimentally infected with *Escherichia coli*." Pp 277-280 in K. Sejrsen, M. Vestergaard, and A. Neimann-Sorensen (eds). *Use of Somatotropin in Livestock Production*. New York: Elsevier Applied Science.

Butterwick, R. F., P. Rowlinson, T. E. C. Weekes, D. S. Parker, and D. G. Armstrong. 1988. "The effect of long-term daily administration ofbovine somatotropin on the daily performance of dairy heifers during their first lactation." *Animal Production* 46:483. (Abstract)

Capen, C. C., and S. L. Martin. 1989. "The thyroid gland." Pp 58-91 in L. E. McDonald (ed). *Veterinary Endocrinology and Reproduction*. Philadelphia, PA: Lea and Febiger.

Capuco, A. V., J. E. Keys, and J. J. Smith. 1989. "Somatotropin increases thyroxine-5'-monodeiodinase activity in lactating mammary tissue of the cow." *Journal of Endocrinology* 121:205-211.

Chalupa, W., L. Baird, C. Soderholm, D. L. Palmquist, R. Hemken, D. Otterby, R. Annexstad, B. Vecchiarelli, R. Harmon, A. Sinha, J. Linn, W. Hansen, F. Ehle, P. Schneider, and R. Eggert. 1987. "Responses of dairy cows to somatotropin." *Journal of Dairy Science* 70(Suppl):176. (Abstract)

Chalupa, W., A. Kutches, D. Swager, T. Lehenbauer, B. Vecchiarelli, R. Shaver, and E. Robb. 1988. "Responses of cows in a commercial dairy to somatotropin." *Journal of Dairy Science* 71(Suppl):210. (Abstract)

Chalupa, W., A. Kutches, D. Swager, T. Lehenbauer, B. Vecchiarelli, R. Shaver, E. Robb, and D. Rock. 1989. "Effects of supplemental somatotropin for two lactations: responses of cows in a commercial dairy." *Journal of Dairy Science* 72(Suppl):327. (Abstract)

Cleale, R. M., J. D. Rehman, E. Robb, A. Sinha, F. R. Ehle, and D. K. Nelson. 1989. "On-farm lactational and reproductive resonses to daily injections of recombinant bovine somatotropin." *Journal of Dairy Science* 72(Suppl):429. (Abstract)

Cole, W. J., P. J. Eppard, G. M. Lanza, R. L. Hintz, K. S. Madsen, S. E. Franson, T. C. White, W. E. Ribelin, B. G. Hammond, S. C. Bussen, R. K. Leak, and L. E. Metzger. 1988. "Response of lactating dairy cows to multiple injections of sometribove, USAN (recombinant methionyl bovine somatotropin) in a prolonged release system. Part II. Health and reproduction." *Journal of Dairy Science* 71(Suppl):184. (Abstract)

Cole, W. J., S. E. Franson, R. G. Hoffman, V. K. Meserole, D. M. Sprick, K. S. Madsen, G. F. Hartnell, D. E. Bauman, H. H. Head, J. T. Huber, and R. C. Lamb. 1989. "Response of cows throughout lactation to sometribove, recombinant methionyl bovine somatotropin, in a prolonged release system--a dose titration study. Part II. Health and reproduction." *Journal of Dairy Science* 72(Suppl):451-452. (Abstract)

Collier, R. J., R. Li, H. D. Johnson, B. A. Becker, F. C. Buonomo, and K. J. Spencer. 1990. "Effect of sometribove (methionyl bovine somatotropin, BST) on plasma insulin-like growth factor I (IGF-I) and II (IGF-II) in cattle exposed to heat and cold stress." *Journal of Dairy Science* 73(Suppl):228. (Abstract)

Davila, D. R., S. Brief, J. Simon, R. E. Hammer, R. L. Brinster, and K. W. Kelley. 1987. "Role of growth hormone in regulating T-dependent immune events in aged, nude, and transgenic rodents." Pp 108-116 in J. R. Perez-Polo, K. Bulloch, R. H. Angeletti, G. A. Hashim, and J. De Vellis (eds). *Neuroimmunomodulation*. New York, NY: Alan R. Liss, Inc.

Davis, S. R., R. J. Collier, J. P. McNamara, and H. H. Head. 1983. Effect of growth hormone and thyroxine treatment of dairy cows on milk production, cardiac output and mammary blood flow. *Proceedings of the New Zealand Society for Endocrinology* 26(Suppl 2):31.

Davis, S. R., P. D. Gluckman, I. C. Hart, and H. V. Henderson. 1987. "Effects of injecting growth hormone or thyroxine on milk production and blood plasma concentrations of insulin-like growth factors I and II in dairy cows." *Journal of Endocrinology* 114:17-24.

de Boer, G., and J. J. Kennelly. 1988. "Effect of somatotropin injection and dietary protein concentration on kinetics of hormones in dairy cows." *Journal of Dairy Science* 71(Suppl):168. (Abstract)

de Boer, G., and J. J. Kennelly. 1989. "Sustained-release bovine somatotropin for dairy cows." *Journal of Dairy Science* 72(Suppl):432. (Abstract)

Dhiman, T., J. Kleinmans, H. D. Radloff, N. J. Tessmann, and L. D. Satter. 1988. "Effect of recombinant bovine somatotropin on feed intake, dry matter digestibility and blood constituents in lactating dairy cows." *Journal of Dairy Science* 71(Suppl):121. (Abstract)

Elvinger, F., P. J. Hansen, H. H. Head, and R. P. Natzke. 1990. "Effect of bovine somatotropin and diet on activity of bovine polymorphonuclear leukocytes and lymphocytes cultured at 38.5 and 42 degrees C." *Journal of Dairy Science* 73:9. (Abstract)

Elvinger, F., H. H. Head, C. J. Wilcox, and R. P. Natzke. 1987. "Effects of administration of bovine somatotropin on lactation milk yield and composition." *Journal of Dairy Science* 70(Suppl):121. (Abstract)

Eppard, P. J., D. E. Bauman, C. R. Curtis, H. N. Erb, G. M. Lanza, and M. J. DeGeeter. 1987. "Effect of 188-day treatment with somatotropin on health and reproductive performance of lactating dairy cows." *Journal of Dairy Science* 70:582-591.

Eppard, P. J., J. L. Vicini, W. J. Cole, and R. J. Collier. 1989. "Effect of Bovine Somatotropin on Animal Health." Pp 74-79 in Proceedings Maryland Nutrition Conference for Feed Manufacturers. College Station, MD: University of Maryland, Department of Poultry and Animal Science.

Esievo, K. A. N., and W. E. Moore. 1979. "Effects of dietary protein and stage of lactation on the hematology and erythrocyte enzymes activities of high-producing dairy cattle." *Research in Veterinary Science* 26:53.

Estrada, J. M., P. G. Reddy, J. E. Shirley, R. A. Frey, and F. Blecha. 1989. "Effect of recombinant bovine somatotropin (rBST) on the immune response of dairy cows." *Journal of Dairy Science* 73(Suppl):150. (Abstract)

Estrada, J. M., and J. E. Shirley. 1990. "Effect of parturition and recombinant bovine somatotropin (rBST) on the metabolic profile of dairy cows." *Journal of Dairy Science* 73(Suppl):199. (Abstract)

Ferguson, J. D., and A. Skidmore. 1989. "Bovine Somatotropin-Reproduction and Health." Pp 57-69 in Proceedings of Advanced Technologies Facing the Dairy Industry: BST. Rochester, NY.

Franson, S. E., W. J. Cole, R. G. Hoffman, V. K. Meserole, D. M. Sprick, K. S. Madsen, G. F. Hartnell, D. E. Bauman, H. H. Head, J. T. Huber, and R. C. Lamb. 1989. "Response of cows throughout lactation to sometribove, recombinant methionyl bovine somatotropin, in a prolonged release system--a dose titration study." *Journal of Dairy Science* 72(Suppl):451. (Abstract)

Fullerton, F. M., I. R. Fleet, R. B. Heap, I. C. Hart, and T. Ben-Mepham. 1989. "Cardiovascular responses and mammary substrate uptake in Jersey cows treated with pituitary-derived growth hormone during late lactation." *Journal of Dairy Research* 56:27-35.

Furniss, S. J., A. J. Stroud, A. C. G. Brown, and G. Smith. 1988. "Milk production, food intakes and weight change of autumn-calving, flat-rate-fed dairy cows given 2-weekly injections of recombinantly-derived bovine somatotropin (BST)." *Animal Production* 46:483. (Abstract)

Gallo, G. F., and E. Block. 1990. "Effects of recombinant bovine somatotropin on nutritional status and liver function of lactating dairy cows." *Journal of Dairy Science* 73:3276-3286.

Galton, D. M. 1989. "Evaluation of sometribove, USAN (recombinant methionyl) bovine somatotropin on milk production health." *Journal of Dairy Science* 72(Suppl):450. (Abstract)

Gibson, J. P., M. van der Meulen, B. W. McBride, and J. H. Burton. 1990. "The effect of rBST administration on fertility and culling rates of lactating dairy cattle." *Journal of Dairy Science* 73:197. (Abstract)

Glimm, D. R., V. E. Baracos, and J. J. Kennelly. 1988. "Effect of bovine somatotropin on the distribution of immunoreactive insulin-like growth factor-I in lactating bovine mammary tissue." *Journal of Dairy Science* 71:2923-2935.

Grings, E. E., D. M. de Avila, R. G. Eggert, and J. J. Reeves. 1990. "Conception rate, growth, and lactation of dairy heifers treated with recombinant somatotropin." *Journal of Dairy Science* 73:73-77.

Hansen, W. P., D. E. Otterby, J. G. Linn, J. F. Anderson, and R. G. Eggert. 1989. "Multi-farm use of bovine somatotropin (BST) and its effects on lactation, health and reproduction." *Journal of Dairy Science* 72(Suppl):429-430. (Abstract)

Hard, D. L., W. J. Cole, S. E. Franson, W. A. Samuels, D. E. Bauman, H. N. Erb, J. T. Huber, and R. C. Lamb. 1988. "Effect of long term sometribove, USAN (recombinant methionyl bovine somatotropin) treatment in a prolonged release system on milk yield, animal health and reproductive performance--pooled across four sites." *Journal of Dairy Science* 71(Suppl):210. (Abstract)

Hemken, R. W., R. J. Harmon, W. J. Silvia, G. Heersche, and R. G. Eggert. 1987. "Response of lactating cows to a second year of recombinant bovine somatotropin (BST) when fed two energy concentrations." *Journal of Dairy Science* 71(Suppl):122. (Abstract)

Heyneman, R., C. Burvenich, M. Van Hoegaerden, and G. Peeters. 1989. "Influence of recombinant bovine somatotropin (rBST) on blood neutrophil respiratory burst activity in healthy cows." *Journal of Dairy Science* 72(Suppl):349. (Abstract)

Huber, J. T. 1987. "The Production Response of BST: Feed Additives, Heat Stress and Injection Intervals." Pp 57-59 in National Invitational Workshop on BST. St. Louis, MO.

Huber, J. T., S. Willam, K. Marcus, and C. B. Theurer. 1988. "Effect of sometribove (SB), USAN (recombinant methionyl bovine somatotropin) injected in lactating cows at 14-d intervals on milk yields, milk composition and health." *Journal of Dairy Science* 71(Suppl):207. (Abstract)

Hutchison, C. F., J. E. Tomlinson, and W. H. McGee. 1986. "The effects of exogenous recombinant or pituitary extracted growth hormone on performance of dairy cows." *Journal of Dairy Science* 69(Suppl):152. (Abstract)

Jenny, B. F., J. E. Ellers, R. B. Tingle, M. Moore, L. W. Grimes, and D. W. Rock. 1988. "Response of dairy cows to recombinant bovine somatotropin in a sustained-release vehicle." *Journal of Dairy Science* 71(Suppl):209. (Abstract)

Johnson, H. D., R. Li, W. Manalu, K. J. Spencer-Johnson, B. A. Becker, R. J. Collier, and C. A. Baile. 1991. "Effects of somatotropin on milk yield and physiologic responses during summer farm and hot laboratory conditions." *Journal of Dairy Science* 74:1250-1262.

Johnsson, I. D., and I. C. Hart. 1986. "Manipulation of milk yield with growth hormone," Pp 105-123 in W. Haresign and D. J. A. Cole (eds). *Recent Advances in Animal Nutrition*. Toronto: Butterworths.

Juskevich, J. C., and C. G. Guyer. 1990. "Bovine growth hormone: human food safety evaluation." *Science* 249:875-884.

Kradel, D. C., R. S. Adams, G. A. Jung, S. B. Guss, W. L. Stout, and C. G. Smiley. 1975. "Blood profiling in cattle--the Pennsylvania experience." *Proceedings of the American Association of Veterinary Laboratory Diagnosticians* 18:327-351.

Kronfeld, D. S. 1965. "Growth hormone-induced ketosis in the cow." *Journal of Dairy Science* 48:342.

Lamb, R. C., M. J. Anderson, S. L. Henderson, J. W. Call, R. J. Callan, D. L. Hard, and L. Kung, Jr. 1988. "Production response of Holstein cows to sometribove USAN (recombinant methionyl bovine somatotropin) in a prolonged release system for one lactation." *Journal of Dairy Science* 71(Suppl):208. (Abstract)

Lanza, G. M., P. J. Eppard, M. A. Miller, S. E. Franson, S. Ganguli, R. L. Hintz, B. G. Hammond, S. C. Bussen, R. K. Leak, and L. E. Metzger. 1988. "Response of lactating dairy cows to multiple injections of sometribove, USAN (recombinant methionyl bovine somatotropin) in a prolonged release system. Part III. Changes in circulating analytes." *Journal of Dairy Science* 71(Suppl):184. (Abstract)

Larson, B. L. (ed.). 1985. *Lactation*. Ames, IA: Iowa State University Press.

Lean, I. J., L. D. Weaver, M. L. Bruss, H. F. Troutt, W. J. Goodger, and J. C. Galland. 1990. "Bovine somatotropin: Management and industry implications." *Compendium on Continuing Education* 12(8):1150-1158.

Li, R., H. D. Johnson, B. A. Becker, K. J. Spencer, and R. J. Collier. 1988. "Effect of cold temperatures on performance of cows supplemented with sometribove (methionyl bovine somatotropin, BST)." *Journal of Dairy Science* 71(Suppl):124. (Abstract)

Lissemore, K. D., K. E. Leslie, B. W. McBride, J. H. Burton, A. R. Willan, and K. G. Bateman. 1991. "Observations on intramammary infection and somatic cell counts in cows treated with recombinant bovine somatotropin." *Canadian Journal of Veterinary Research* 55:196-198.

Lormore, M. J., L. D. Muller, D. R. Deaver, and L. C. Griel, Jr. 1990. "Early lactation responses of dairy cows administered bovine somatotropin and fed diets high in energy and protein." *Journal of Dairy Science* 73:3237-3247.

Marcek, J. M., W. J. Seaman, and J. L Nappier. 1989. "Effects of repeated high dose administration of recombinant bovine somatotropin in lactating dairy cows." *Veterinary and Human Toxicology* 31:455-460.

Mattos, M., A. V. Pires, V. P. de Faria, J. A. Duque, and K. S. Madsen. 1989. "The effect of sometribove (recombinant methionyl bovine somatotropin) on milk yields and composition in lactating dairy cows in Brazil." *Journal of Dairy Science* 72(Suppl):452. (Abstract)

McBride, B. W., J. L. Burton, and J. H. Burton. 1988. "The influence of bovine growth hormone (somatotropin) on animals and their products." *Research and Development in Agriculture* 5(1):1-21.

McClean, C. and B. Laarveld. 1988. "Effect of somatotropin treatment and type of protein supplement on thyroid function of dairy cows." *Journal of Dairy Science* 71(Suppl):122. (Abstract)

McDaniel, B. T., W. E. Bell, J. Fetrow, B. D. Harrington, and J. D. Rehman. 1990. "Survival rates and reasons for removal of cows injected with rbST." *Journal of Dairy Science* 73(Suppl):159. (Abstract)

McDaniel, B. T., D. M. Gallant, J. Fetrow, B. Harrington, W. E. Bell, P. Hayes, and J. D. Rehman. 1989. "Lactational, reproductive and health responses to recombinant bovine somatotropin under field conditions." *Journal of Dairy Science* 72(Suppl):429. (Abstract)

McGuffey, R. K., R. P. Basson, D. L. Snyder, E. Block, J. H. Harrison, A. H. Rakes, R. S. Emery, and L. D. Muller. 1991. "Effect of somidobove sustained-release administration on the lactation performance of dairy cows." *Journal of Dairy Science* 74:1263-1276.

McGuffey, R. K., H. B. Green, R. P. Basson, and T. H. Ferguson. 1989. "Lactation response of dairy cows receiving bovine somatotropin via daily injections or in a sustained-released vehicle." *Journal of Dairy Science* 73:763-771.

Mohammed, M. E., and H. D. Johnson. 1985. "Effect of growth hormone on milk yields and related physiological functions of Holsteins exposed to heat stress." *Journal of Dairy Science* 68:1123-1133.

Morbeck, D. E., B. T. McDaniel, and J. H. Britt. 1989. "Reproductive and metabolic performance of primiparous Holstein cows treated with recombinant bovine somatotropin (rbST)." *Journal of Dairy Science* 72(Suppl):345. (Abstract)

Moseley, W. M. 1989. "Bovine somatotropin (bSt) pretreatment does not enhance the GnRH-induced LH response in pre- and postpubertal Holstein heifers." *Journal of Dairy Science* 72(Suppl):344. (Abstract)

Munneke, R. L., J. L. Sommerfeldt, and E. A. Ludens. 1988. "Lactational responses of dairy cows to recombinant bovine somatotropin." *Journal of Dairy Science* 71(Suppl):206. (Abstract)

Murphy, M., D. O'Callaghan, M. Rath, and J. F. Roche. 1989. "The effect of bovine somatotropin with or without avopacin on milk yield, milk composition, body weight, body condition score and reproductive performance of autumn calving Friesian dairy cows." *Journal of Dairy Science* 72(Suppl):444-445. (Abstract)

Palmquist, D. L. 1988. "Response of high-producing cows given daily injections of recombinant bovine somatotropin from D 30-296 of lactation." *Journal of Dairy Science* 71(Suppl):206. (Abstract)

Peel, C. J., and D. E. Bauman. 1987. "Somatotropin and lactation." *Journal of Dairy Science* 70:474-486.

Peel, C. J., D. E. Bauman, R. C. Gorewit, and C. J. Sniffen. 1981. "Effect of exogenous growth hormone on lactational performance in high yielding dairy cows." *Journal of Nutrition* 111:1662-1671.

Pell, A. N., D. S. Tsang, M. T. Huyler, B. A. Howlett, J. Kunkel, and W. A. Samuels. 1988. "Responses of Jersey cows to treatment with sometribove, USAN (recombinant methionyl bovine somatotropin) in a prolonged release system." *Journal of Dairy Science* 71(Suppl):206. (Abstract)

Phipps, R. H. 1989. "A review of the influence of somatotropin on health, reproduction and welfare in lactating dairy cows." Pp 88-119 in K. Sejrsen, M. Vestergaard, and A. Neimann-Sorensen (eds). *Use of Somatotropin in Livestock Production*. New York, NY: Elsevier Applied Science.

Phipps, R. H., R. F. Weller, A. R. Austin, N. Craven, and C. J. Peel. 1988. "A preliminary report on a prolonged release formulation of bovine somatotropin with particular reference to animal health." *Veterinary Record* 122:512-513. (Abstract)

Pocius, P. A., and J. H. Herbein. 1986. "Effects of in vivoin vitro administration of growth hormone on milk production and in vitro hepatic metabolism." *Journal of Dairy Science* 69:713-720.

Prosser, C. G., I. R. Fleet, and A. N. Corps. 1989. "Increased secretion of insulin-like growth factor I into milk of cows treated with recombinantly-derived bovine growth hormone." *Journal of Dairy Research* 56:17-26.

Rajamahendran, R., S. Desbottes, J. A. Shelford, R. G. Peterson, and J. J. Kennelly. 1989. "Effect of recombinant bovine somatotropin (rbSt) on milk production and reproductive performance of dairy cows." *Journal of Dairy Science* 72(Suppl):444. (Abstract)

Rijpkema, Y. S., L. Van Reeuwyk, C. J. Peel, and E. P. Mol. 1987. "Response of dairy cows to long-term treatment with somatotropin in a prolonged release formulation." *Proceedings of the European Association for Animal Production* 38.

Rock, D. W., D. L. Patterson, W. Chalupa, J. H. Clark, R. M. DeGregorio, and B. F. Jenny. 1989. "Lactation performance of dairy cows given a sustained-release form of recombinant bovine somatotropin." *Journal of Dairy Science* 72(Suppl):431. (Abstract)

Roets, E., C. Burvenich, G. Vandeputte-Van Messom, and R. M. Akers. 1988. "*Escherichia coli* induced mastitis in lactating cows: influence of bovine somatotropin on a-lactalbumin concentration in blood and milk." *European Journal of Physiology* 412(Suppl 1):51. (Abstract)

Rowe-Bechtel, C. L., L. D. Muller, D. R. Deaver, and L. C. Griel, Jr. 1988. "Administration of recombinant bovine somatotropin (rbST) to lactating dairy cows beginning at 35 and 70 days postpartum. I. Production response." *Journal of Dairy Science* 71(Suppl):166. (Abstract)

Rowlands, G. J., W. Little, A. J. Stark, and R. Manston. 1979. "The blood composition of cows in commercial dairy herds and its relationships with season and lactation." *British Veterinary Journal* 135:64.

Rowlinson, P., R. F. Butterwick, T. E. C. Weekes, and D. S. Parker. 1989. "The effect of bovine somatotropin on the performance of dairy cows during their first two lactations." *Proceedings of the International Conference on Production Disease in Farm Animals* 7:324-326.

Samuels, W. A., D. L. Hard, R. L. Hintz, P. K. Olsson, W. J. Cole, and G. F. Hartnell. 1988. "Long term evaluation of sometribove, USAN (recombinant methionyl bovine somatotropin) in a prolonged release system for lactating cows." *Journal of Dairy Science* 71(Suppl):209. (Abstract)

Schalm, O. W., N. C. Jain, and E. J. Carroll. 1975. *Veterinary Hematology*, 3rd Ed. Philadephia, PA: Lea and Febiger.

Schams, D., B. Graule, M. Thyerl-Abele, F. Graf, and C. Wollny. 1989. "Insulin-like growth factor I and somatotropin during lactation and after treatment with sometribove (recombinant methionyl BST) in German Fleckvieh and German Black and White cows." *Journal of Dairy Science* 72(Suppl):347. (Abstract)

Schemm, S. R., D. R. Deaver, L. C. Griel, Jr., and L. D. Muller. 1990. "Effects of recombinant bovine somatotropin on luteinizing hormone and ovarian function in lactating dairy cows." *Biology of Reproduction* 42:815-821.

Sechen, S. J., F. R. Dunshea, and D. E. Bauman. 1990. "Somatotropin in lactating cows: effect on response to epinephrine and insulin." *American Journal of Physiology* 258:E582-E588.

Sechen, S. J., S. N. McCutcheon, and D. E. Bauman. 1989. "Response to metabolic challenges in early lactation dairy cows during treatment with bovine somatotropin." *Domestic Animal Endocrinology* 6(2):141-154.

Soderholm, C. G., D. E. Otterby, J. G. Linn, F. R. Ehle, J. E. Wheaton, W. P. Hansen, and R. J. Annexstad. 1988. "Effects of recombinant bovine somatotropin on milk production, body composition, and physiological parameters." *Journal of Dairy Science* 71:355-365.

Spicer, L. J., W. B. Tucjer, and G. D. Adams. 1990. "Insulin-like growth factor-I in dairy cows: relationships among energy balance, body condition, ovarian activity, estrous behavior." *Journal of Dairy Science* 73:929-937.

Staples, C. R., and H. H. Head. 1988. "Short-term administration of bovine somatotropin to lactating dairy cows in a subtropical environment." *Journal of Dairy Science* 71:3274-3282.

Staples, C. R., C. R. Thatcher, and J. H. Clark. 1990. "Relationship between ovarian activity and energy status during the early postpartum period of high producing dairy cows." *Journal of Dairy Science* 73(Suppl):938-947.

Tanner, J. W., and S. D. Hauser. 1989. "Molecular evidence for the presence of the somatotropin receptor in the bovine ovary." *Journal of Dairy Science* 72(Suppl):413. (Abstract)

Thomas, C., I. D. Johnsson, W. J. Fisher, G. A. Bloomfield, and S. V. Morant. 1987a. "The effect of recombinant bovine somatotropin on milk production, reproduction and health of dairy cows." *Animal Production* 44:460-461. (Abstract)

Thomas, C., I. D. Johnsson, W. J. Fisher, G. A. Bloomfield, S. V. Morant, and J. M. Wilkinson. 1987b. "Effect of somatotropin on milk production, reproduction and health of dairy cow." *Journal of Dairy Science* 70(Suppl):175. (Abstract)

Thomas, J. W., R. A. Erdman, D. M. Galton, R. C. Lamb, M. J. Arambel, J. D. Olson, K. S. Madsen, W. A. Samuels, C. J. Peel, and G. A. Green. 1991. "Responses by lactating cows in commercial dairy herds to recombinant bovine somatotropin." *Journal of Dairy Science* 74:945-964.

Thomas, J. W., W. A. Samuels, and K. S. Madsen. 1989. "Use of sometribove, USAN (recombinant methionyl) in a prolonged release system in commercial dairy herds." *Journal of Dairy Science* 72(Suppl):450. (Abstract)

Vicini, J. L., J. M. DeLeon, W. J. Cole, P. J. Eppard, G. M. Lanza, S. Hudson, and M. A. Miller. 1988. "Effect of acute administration of extremely large doses of sometribove, USAN (recombinant methionyl bovine somatotropin) in a prolonged release formulation on milk production and health of dairy cows." *Journal of Dairy Science* 71(Suppl):168. (Abstract)

Vicini, J. L., S. Hudson, W. J. Cole, M. A. Miller, P. J. Eppard, T. C. White, and R. J. Collier. 1990. "Effect of acute challenge with an extreme dose of somatotropin in a prolonged-release formulation on milk production and health of dairy cattle." *Journal of Dairy Science* 73:2093-2102.

Weller, R. F., R. H. Phipps, N. Craven, and C. J. Peel. 1990. "Use of prolonged-release bovine somatotropin for milk production in British Friesian dairy cows." *Journal of Agricultural Science* 115:105-112.

West, J. W., K. Bondari, J. C. Johnson, Jr., K. A. Ash, and V. N. Taylor. 1989. "The response of lactating Holstein and Jersey cows to recombinant bovine somatotropin (rbST) administered during hot, humid weather." *Journal of Dairy Science* 72(Suppl):427-428. (Abstract)

West, J. W., J. C. Johnson, Jr. , and K. Bondari. 1988. "The effect of bovine somatotropin on productivity and physiologic responses of lactating Holstein and Jersey cows." *Journal of Dairy Science* 71(Suppl):209. (Abstract)

West, J. W., B. G. Mullnix, J. C. Johnson, Jr., K. A. Ash, and V. N. Taylor. 1990. "Effects of bovine somatotropin on dry matter intake, milk yield, and body temperature in Holstein and Jersey cows during heat stress." *Journal of Dairy Science* 73:2896-2906.

West, J. W., B. G. Mullnix, and T. G. Sandifer. 1991. "Effects of bovine somatotropin on physiologic responses of lactating Holstein and Jersey cows during hot, humid weather." *Journal of Dairy Science* 74:840-851.

Whitaker, D. A., E. J. Smith, and J. M. Kelly. 1989. "Milk production, weight changes and blood biochemical measurements in dairy cattle receiving recombinant bovine somatotropin." *Veterinary Record* 124(4):83-86.

Whitaker, D. A., E. J. Smith, J. M. Kelly, and L. S. Hodgson-Jones. 1988. "Health, welfare and fertility implications of the use of bovine somatotropin in dairy cattle." *Veterinary Record* 122:503-505. (Abstract)

White, T. C., G. M. Lanza, S. E. Dyer, S. Hudson, S. E. Franson, R. L. Hintz, J. A. Duque, S. C. Bussen, R. K. Leak, and L. E. Metzger. 1988. "Response of lactating dairy cows to intramuscular or subcutaneous injection of sometribove, USAN (recombinant methionyl bovine somatotropin) in a 14-day prolonged release system. I. Animal performance and health." *Journal of Dairy Science* 71(Suppl):167. (Abstract)

Wilfond, D. H., K. C. Bachman, H. H. Head, C. J. Wilcox, and G. F. Hartnell. 1989. "Effect of dry period administration of sometribove upon lactational performance of Holstein cows." *Journal of Dairy Science* 72(Suppl):329. (Abstract)

Zoa-Mboe, A., H. H. Head, K. C. Bachman, F. Baccari, Jr., and C. J. Wilcox. 1989. "Effects of bovine somatotropin on milk yield and composition, dry matter intake, and some physiological functions of holstein cows during heat stress." *Journal of Dairy Science* 72:907-916.

Chapter 5
Appendix

TABLE 5.A. BST Trials With Dairy Animals in Which Some Health-related Events Were Evaluated, by Number of Cows, Days in Milk at Start, Dose, and Duration.

Authors	*Trial Group #*	*Group Size (cows)*	*Time at Start (DIM)*[a]	*Dose (mg)*	*Duration (days)*
Eppard et al. 1987	1	6	84±10	0.0	188
	2	6	84±10	13.5/d	188
	3	6	84±10	27.0/d	188
	4	6	84±10	40.5/d	188
Soderhold et al. 1988	1	7	28<35	0.0	266
	2	7	28<35	10.3/d	266
	3	7	28<35	20.6/d	266
	4	7	28<35	41.2/d	266
Bauman et al. 1989	1	40	60±3	0.0	>252
	2	40	60±3	500/14d	>252
Zoa-Mboe et al. 1989	1	13	46<106	0.0	29
	2	13	46<106	20.6/d	29
West et al. 1990	1	17	182 (avg)	0.0	80
	2	17	182 (avg)	20/d	80
Thomas et al. 1987a	1	20	~35	0.0	294
	2	20	~35	12.5/d	294
	3	20	~35	25/d	294
	4	20	~35	50/d	294
Burton, J. H. et al. 1987	1	9	28<35	0.0	266
	2	10	28<35	12.5/d	266
	3	10	28<35	25/d	266
	4	9	28<35	50/d	266

(continues)

[a] DIM = Days-in-milk (days postpartum)

TABLE 5.A. *(continued)*

Authors	*Trial Group #*	*Group Size (cows)*	*Time at Start (DIM)*[a]	*Dose (mg)*	*Duration (days)*
Chalupa et al. 1987	1	34	28<35	0.0	?
	2	34	28<35	12.5/d	
	3	34	28<35	25/d	
	4	34	28<35	50/d	
		(9 cows)			
Hemken et al. 1987	1	10	28<35	0.0	~270
	2	10	28<35	20.6/d 2nd yr	~270
	3	10	28<35	had BST 1st yr but not 2nd yr	~270
Johnson et al. 1991	1	6	?	0.0	30
	2	6		?	30
Rowe-Bechtel et al. 1988	1	16	35<70	0.0	200 PP
	2	16	35<70	25/d	200 PP
Cole et al. 1988	1	~20	60	1x600/14d	~245
	2	~20	60	3x600/14d	~245
	3	~20	60	5x600/14d	~245
	4	~20	60	0.0	~245
Palmquist 1988	1	9	28<35	0.0	~266
	2	9	28<35	10.3/d	~266
	3	9	28<35	20.6/d	~266
	4	9	28<35	41.2/d	~266
Peel et al. 1981	1	22	60±3	0.0	238
	2	23	60±3	500/14d	238

(continues)

[a] DIM = Days-in-milk (days postpartum)

TABLE 5.A. *(continued)*

Authors	*Trial Group #*	*Group Size (cows)*	*Time at Start (DIM)*[a]	*Dose (mg)*	*Duration (days)*
Munneke et al.	1	8	75±15	0.0	230
1988	2	8	75±15	5/d	230
	3	8	75±15	10/d	230
	4	8	75±15	15/d	230
	5	8	75±15	20/d	230
Huber et al.	1	40	60±3	0.0	252
1988	2	40	60±3	500/14d	252
Lamb et al.	1	36	60±3	0.0	~250
1988	2	36	60±3	500/14d	~250
Aguilar et al.	1	49	?	0.0	112
1988	2	48		25/d	112
Jenny et al.	1	9	98<105	0.0	182
1988	2	9	98<105	140/14d	182
	3	9	98<105	350/14d	182
	4	9	98<105	700/14d	182
Samuels et al.	1	63	60±3	0.0	252
1988	2	63	60±3	500/14d	252
West et al.	1	8	75±7	0.0	~230
1988	2	8	75±7	5/d	~230
	3	8	75±7	10/d	~230
	4	8	75±7	15/d	~230
	5	8	75±7	20/d	~230
Hard et al.	1	182	60±3	0.0	252
1988	2	182	60±3	500/14d	252

(continues)

[a] DIM = Days-in-milk (days postpartum)

TABLE 5.A. *(continued)*

Authors	*Trial Group #*	*Group Size (cows)*	*Time at Start (DIM)*[a]	*Dose (mg)*	*Duration (days)*
Chalupa et al.	1	30	~98	0.0	203
1988	2	30	~98	10.3/d	203
	3	30	~98	20.6/d	203
	4	30	~98	30.9/d	203
Bunn et al.	1	~66	95<109	0.0	~200
1988	2	66	95<109	350/14d	~200
	3	66	95<109	560/14d	~200
Chalupa et al.	1	15	?	0.0	through
1989	2	15		10.3/d	2nd
	3	15		20.6/d	lactation
Wilford et al.	1	68	Dry period	25/d	14
1989	2	67	Dry period	0.0	14
Schams et al.	1	?	Variable	0.0	Variable
1989	2	?	Variable	500/14d	Variable
Cleale et al.	1	~156	98<105	0.0	~210
1989	2	155	98<105	20.6/d	~210
	3	155	98<105	41.2/d	~210
McDaniel et al.	1	~51	~28	0.0	~275
1989	2	51	~28	5.15/d	~275
	3	51	~28	10.3/d	~275
	4	51	~28	16.5/d	~275
Hansen et al.	1	~88	28<35	0.0	273
1989	2	88	28<35	5.15/d	273
	3	88	28<35	10.3/d	273
	4	88	28<35	16.5/d	273

(continues)

[a] DIM = Days-in-milk (days postpartum)

TABLE 5.A. *(continued)*

Authors	*Trial Group #*	*Group Size (cows)*	*Time at Start (DIM)*[a]	*Dose (mg)*	*Duration (days)*
Rock et al. 1989	1	~37	98<105	0.0	~200
	2	37	98<105	140/14d	~200
	3	37	98<105	350/14d	~200
	4	37	98<105	700/14d	~200
de Boer and Kennelly 1989	1	~10	98<112	0.0	196
	2	10	98<112	10.3/d	196
	3	10	98<112	175/d	196
	4	10	98<112	350/14d	196
Rajamahendran et al. 1989	1	36	28<35	0.0	270
	2	36	28<35	10.3/d	270
	3	36	28<35	20.6/d	270
Thomas, J. W. et al. 1989	1	133	Varied	0.0	84
	2	133	Varied	500.0/14d	84
Galton et al. 1989	1	~98	57<189	500/14d	84
	2	98	57<189	0.0	84
Arambel et al. 1989	1	~54	57<180	0.0	84
	2	54	57<180	500/14d	84
Cole et al. 1989	1	~64	60±3	0.0	~250
	2	64	60±3	250/14d	~250
	3	64	60±3	500/14d	~250
	4	64	60±3	700/14d	~250
Mattos et al. 1989	1	~30	60<180	0.0	112
	2	30	60<180	500/14d	112

(continues)

[a] DIM = Days-in-milk (days postpartum)

TABLE 5.A. *(continued)*

Authors	*Trial Group #*	*Group Size (cows)*	*Time at Start (DIM)[a]*	*Dose (mg)*	*Duration (days)*
Marcek et al.	1	2	~150	0.0	21
1989	2	2	~150	430/d	21
Schemm et al.	1	8	35	0.0	70
1990	2	8	70	0.0	70
	3	8	35	25/d	70
	4	8	70	25/d	70
Capuco et al.	1	6	non-preg	0.0	5
1989	2	6	non-preg	40/d	5
Whitaker et al.	1	20	80±7	0.0	14
1989	2	20	80±7	500/14d	14
Weller et al.	1	45	?	0.0	~244 2nd lactation
1990					
	2	45	?	500/14 d	~244 2nd lactation
Burvenich et al.	1	7	21<63	0.0	10
1988	2	7/10	21<63	40/d	10
McGuffey et al.	1	8	44<138	0.0	14
1989	2	8	44<138	25.0/d[b]	14
No. 1:	3	8	44<138	25.0/d[c]	14
No. 2:	1	7	63<246	0.0	84
	2	7	63<246	960.0/14d	84

(continues)

[a] DIM = Days-in-milk (days postpartum)
[b] Sephadex
[c] Tris-acryl

TABLE 5.A. *(continued)*

Authors	*Trial Group #*	*Group Size (cows)*	*Time at Start (DIM)*[a]	*Dose (mg)*	*Duration (days)*
No. 3:	1	7	?	0.0	84
	2	7		320.0/14d	84
	3	7		640.0/21d	84
	4	7		960.0/28d	84
Vicini et al. 1988	1	8	Pregnant	30g total dose	14
White et al. 1988	1	32	~60	0.0	~250
	2	32	~60	500/14d	~250
Morbeck et al. 1989	1	8	28<35	0.0	~270
	2	8	28<35	6.25/d	~270
	3	8	28<35	12.5/d	~270
	4	8	28<35	20/d	~270
Estrada et al. 1989	1	8	90	0.0	28
	2	8	90	25/d	28
Burton, J. L. et al. 1989a	1	12	28	0.0	294
	2	10	28	10.3/d	294
	3	10	28	20.6/d	294
Burton, J. L. et al. 1989b	1	15	28<35	0.0	266
	2	15	28<35	10.3/d	266
	3	13	28<35	20.6/d	266
Furniss, et al. 1988	1	12	57<65	0.0	280
	2	12	57<65	500/14d	280
Butterwick et al. 1988	1	7	14	0.0	280
	2	6	14	25/d	280
	3	6	70	25/d	266

(continues)

[a] DIM = Days-in-milk (days postpartum)

TABLE 5.A. *(continued)*

Authors	*Trial Group #*	*Group Size (cows)*	*Time at Start (DIM)*[a]	*Dose (mg)*	*Duration (days)*
Lormore et al. 1990	1	9	25	0.0[d]	125
	2	9	25	25.0/d[e]	125
	3	9	25	0.0[f]	125
	4	9	25	25.0/d[g]	125
Dhiman et al. 1988	1	32	91	0.0	210
	2	32	91	20.6/d	210
Hutchison et al. 1986	1	~8	84±10	0.0/d	188
	2	8	84±10	13.5/d	188
	3	8	84±10	27.0/d	188
	4	8	84±10	40.5/d	188
Elvinger et al. 1987	1	9	28<35	0.0	273
	2	9	28<35	6.25/d	273
	3	9	28<35	12.5/d	273
	4	9	28<35	25.0/d	273
Sechen et al. 1989	1	4	61±2	0.0	14
	2	4	61±2	26.3/d	14
Peel et al. 1981	1	5	74	0.0	11
	2	5	74	51.5Iμ/d	11
Prosser et al. 1989	1	6	245<329	30/d	7
		(non-pregnant)			

(continues)

[a] DIM = days-in-milk (days postpartum)
[d] Control diet
[e] Control diet + BST
[f] High-density diet w/o BST
[g] High-density diet plus BST

TABLE 5.A. *(concluded)*

Authors	*Trial Group #*	*Group Size (cows)*	*Time at Start (DIM)*[a]	*Dose (mg)*	*Duration (days)*
Grings et al. 1990	1	40	heifers	41.2/d	150
	2	40	heifers	0.0	150
Gibson et al. 1990	1	~120	28<35	0.0	266
	2	~120	28>35	10.3/d	266
	3	~120	28<35	20.6/d	266
	4	~120	28<35	41.2/d	266
Estrada and Shirley 1990	1	16	90	25.0/d	28

[a] DIM = days-in-milk (days postpartum)

PART THREE

Bovine Somatotropin and the Dairy Sector

6

Economic Evaluation of BST for Onfarm Use

L. J. (Bees) Butler

The essence of an evaluation of BST for onfarm use is to examine the net revenue associated with its use. Ideally, a full analysis would be carried out, with a budgeting exercise to assess the full implications of using BST in a whole-farm context. The following analysis is a very basic way of carrying out an initial evaluation of BST on the farm. Essentially, it applies some shortcut methods of analysis to the dairy operation, with the aim of assessing if BST is a feasible technology to adopt, and for which cows.

The analysis can be formed by hand calculations, but using a computer spreadsheet program (such as Lotus 1-2-3, Quattro Pro, Excel, Supercalc), is desirable because even the initial evaluation involves an assessment of BST feasibility for each cow in the herd. There are still many unknowns with respect to response rates, price received for milk, and cost of labor, feed, BST, and other inputs. Thus, a thorough analysis of BST should involve a "best-guess scenario" as well as a "worst-case scenario."

Basic Analysis

Change in Gross and Net Revenue

Net revenues generated from the use of BST equal the change in gross revenues from using BST minus the change in costs associated with using BST, i.e.,

$$NR = \Delta GR - \Delta TC,$$

where NR = net revenues associated with using BST
ΔGR = change in gross revenues, and
ΔTC = change in total costs.

The change in gross revenues is calculated by multiplying the production base of the cow times the response rate to BST times the milk price, i.e.,

$$\Delta GR = PB \times RR \times MP,$$

where PB = production base of the cow, e.g., 14,000 lb, 20,000 lb, etc.,
RR = response rate of BST, as a percentage of the production base, e.g. 10%, 12%, 15% etc., and
MP = current price ($/cwt) received for milk.

For example, a cow producing 16,800 lb of milk per year with a 12 percent response rate from BST will produce an extra 2,016 lbs of milk per year (16,800 x 0.12) or 20.16 cwt/year. If the price received for milk is currently $11.50/cwt then the change in gross revenue is $231.84 (20.16 x 11.50).

Changes in the Cost of Milk Production

The change in total costs due to BST requires a bit more detail. Assume, for example, the following:

Milk price	=	$11.50/cwt,
Hauling costs	=	$0.20/cwt,
Extra breeding, labor, and veterinary costs	=	$0.47/cwt,
Cost of BST	=	$0.35/cow/day,
Feed-cost increases	=	6%/each 10% increase in milk production,
Response rate of BST	=	12%/lactation, and
Base feed costs	=	201 + 0.03 x production base of cow.

The change in total costs equals the change in feed costs + the cost of BST + change in other costs, i.e.,

$$\Delta TC = \Delta FC + CBST + \Delta OC$$

where ΔFC = change in feed costs,
$CBST$ = cost of BST, and
ΔOC = change in other costs such as veterinary services, breeding, hauling milk, etc.

The change in feed costs is calculated from the assumptions above as an increase of 6 percent for each 10 percent increase in milk production, i.e.,

$$\Delta FC = (RR \times CHFC \times BFC)$$

where RR = response rate,
$CHFC$ = change in feed costs, and
BFC = base feed costs.

Thus, for a cow producing (without BST) 16,800 lb milk/lactation, the change in feed costs associated with BST treatment will be 0.12 x 0.06 x 10 x [201 + (0.03 x 16,800)] = $50.76. That is, it will cost an extra $50.76 in feed costs to get the additional 2,016 lbs of milk from the cow, given the assumptions made.

The cost of BST is unknown at this stage. Most analysts are guessing between 25 and 75 cents per cow per day. Thus,

$$CBST = CHP \times ND$$

where $CBST$ = total cost of BST,
CHP = cost of BST per cow per day, and
ND = number of days BST is administered.

From the example and assumptions, if BST is administered for 215 days (mid and late lactation) at $0.35/cow/day, then the total per-cow cost of BST will be 0.35 x 215 or $75.25.

Estimating the change in other costs associated with using BST is perhaps the most difficult. There may be additional labor costs from having to inject the BST; there may be additional veterinary costs associated with the intensive management required for cows on BST and for possible increases in breeding costs; and there will be increased hauling costs associated with the additional milk produced. The change in other costs is given by

$$\Delta OC = PB \times RR \times VC$$

where PB = production base of the cow in cwt,
RR = response rate, and
VC = increased costs associated with using BST.

From the assumptions previously:

Hauling fees	=	$0.20 per cwt,
Vet costs, etc.	=	$0.47 per cwt, and
Total increase in costs	=	$0.67 per cwt.

Thus, for a cow milking 16,800 lb (168 cwt) and having a response rate of 12 percent, the change in other costs associated with using BST is (168 x 0.12 x 0.67) = $13.51.

Changes in Net Revenues Associated with BST

Under the above assumptions then, the change in total costs associated with BST use are:

Change in feed costs	=	$ 50.76
Cost of BST	=	$ 75.25
Change in other costs	=	$ 13.51
Change in total costs	=	$139.52

Thus, in this case, net revenue from using BST is:

$$\begin{aligned} NR &= \Delta GR - \Delta TC \\ &= \$231.84 - \$139.52 \\ &= \$\ 92.32 \end{aligned}$$

The above analysis is applied to a cow producing (base level) 16,800 lb of milk per year, with a response rate of 12 percent and milk price of $11.50/cwt, together with a change in feed costs of 6 percent/each 10 percent increase in milk production, base feed costs of $(201 + 3% of production base), and where BST costs $0.35 per cow per day. Obviously, net returns to BST will vary depending on the assumptions made.

Figures 6.1, 6.2, and 6.3 present estimates of net returns to BST under various assumptions for cows with production bases of from 12,000 to 21,000 lbs and for response rates from 5 to 20 percent, milk prices from $10 to $13 per cwt, and the cost of BST from $0.20 to

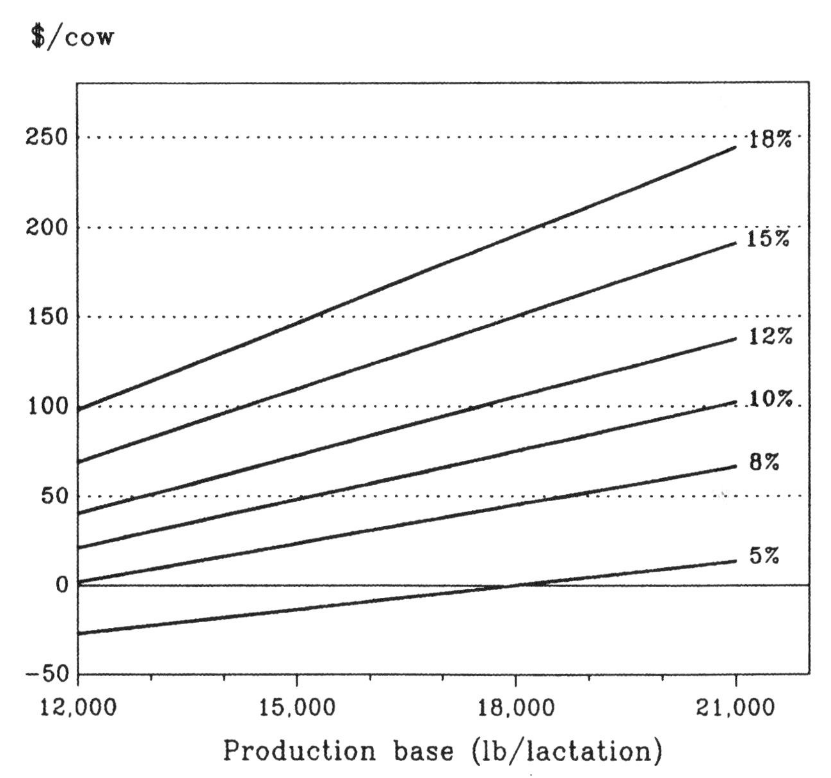

Assumptions:
Milk price: $11.50/cwt
Change in feed costs: 6% per 10% increase in milk production
BST cost: $0.35/cow/day
Additional costs: veterinary, labor, milk-transport, etc: $0.67/additional cwt produced

FIGURE 6.1. Change in net revenue ($/cow/lactation), assuming BST use for 215 days, by six response rates (5 to 18%) and four production bases (12,000 to 21,000 lb/lactation).

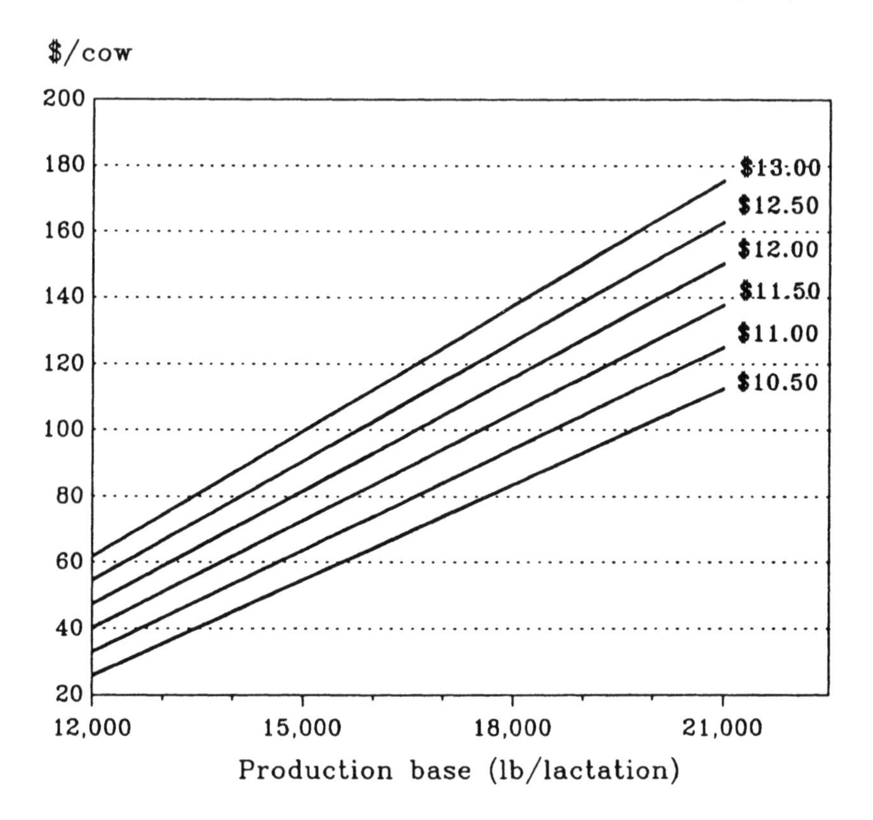

Assumptions:
Response rate: 12%
Change in feed costs: 6% per 10% increase in milk production
BST cost: $0.35/cow/day
Additional costs: veterinary, labor, milk-transport, etc: $0.67/additional cwt produced

FIGURE 6.2. Change in net revenue ($/cow/lactation), assuming BST use for 215 days, by six milk-price levels ($10.50 to 13.00/cwt) and four production bases (12,000 to 21,000 lb/lactation).

$0.80/cow/day. The range of net returns to BST is extremely wide. For example, with a response rate of only 5 percent, BST would not be feasible for cows with base production of less than 18,000 lb, given the assumptions.

Sensitivity Analysis. In order to make timely and well-informed decisions, producers require information not only about the feasibility of adopting a new technology, but also about the relative sensitivity of the parameters that affect the feasibility analysis. While Figures 6.1,

6.2, and 6.3 reflect some sense of the ranges of net returns to BST, it is difficult to judge the relative sensitivity of the parameters to potential change. For example, a cow milking 15,000 lb will return $63.82 in additional net revenues when milk prices are at $11.00 per cwt. But what happens if the response rate of 12 percent is not achieved, and milk prices fall to $10.50/cwt? A quick spreadsheet analysis would indicate that if prices fell to $10.50/cwt, net returns would be $33.14 at a response rate of 10 percent and $11.46 if the response rate is only 8 percent.

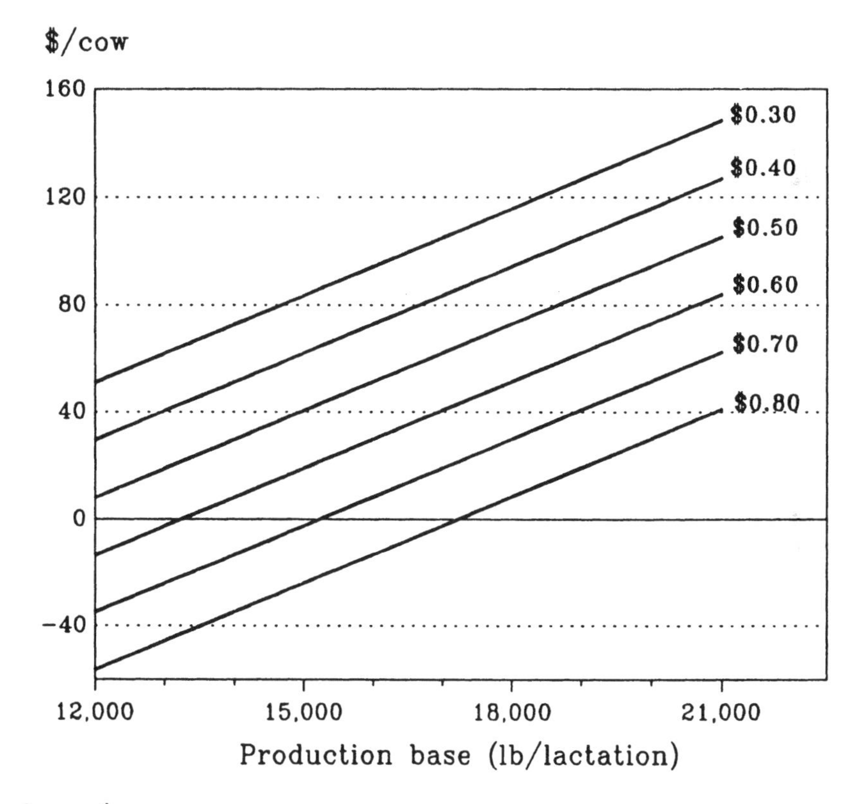

Assumptions:
Response rate: 12%
Milk price: $11.50/cwt
Change in feed costs: 6% per 10% increase in milk production
Additional costs: veterinary, labor, milk-transport, etc: $0.67/additional cwt produced

FIGURE 6.3. Change in net revenue ($/cow/lactation), assuming BST use for 215 days, by six levels of BST cost ($0.30 to 0.80/cow/day) and four production bases (12,000 to 21,000 lb/lactation).

A somewhat more useful way to examine the sensitivity of net returns to changes in response rates, milk prices, etc., is to derive the elasticity of net returns with respect to the variable in question. The elasticity of net returns, with respect to a variable, is defined as the percentage change in net returns due to a 1 percent change in the variable in question. It is defined mathematically as

$$(dNR/dX)(X/NR)$$

where NR = net returns to BST, and
X = variable in question.

Elasticities are typically not constant at different levels of the variables under consideration. However, the ranges of variation in elasticities are much smaller than the ranges of variation in net returns presented in Figures 6.1, 6.2, and 6.3. The elasticity (sensitivity) of net returns to various parameters are given in Table 6.1, with all parameters evaluated at the following levels:

Milk price	$11.50/cwt
Change in feed costs	6.0% per 10% increase in milk production
Cost of BST	$ 0.35/cow/day
Number of days administered	215
Veterinary services and other costs	$ 0.67/additional cwt produced
Response rate	12.0%
Production base	16,800 lbs
Base feed costs	$201 + 0.03 (production base in lb/lactation)
Change in net revenues using the above parameters	$92.32

As suggested above, these elasticities vary considerably with changes in the basic assumptions. However, these elasticities remain proportionately constant to each other, and therefore can be used to indicate the relative sensitivity of net revenues to changes in each of the parameters. For example, the most sensitive parameter appears to be milk price. For each 1 percent increase in the price of milk, net revenues increase 2.51 percent. However, for each 1 percent increase in the cost of BST, net revenues decrease 0.82 percent.

While the elasticities given in Table 6.1 give some indication of the relative sensitivity of net returns to BST due to changes in the various parameters, they must be interpreted with care. For example, a one percent increase in the milk price is 1.15 cents, while a 2.51 percent increase in net revenues is $2.32. Clearly, the relative proportions implied by the elasticities do not give a clear picture of the relative sensitivity of the parameters. Therefore, it is worthwhile to develop "rules of thumb" for analyzing the economic aspects of BST use. The "rules of thumb" given in Table 6.2 are guidelines to the relative sensitivity of net returns to BST due to changes in various parameters. These "rules-of-thumb" are much more useful--and will now allow the casual observer to answer the question posed at the beginning of this section. A $0.50 fall in milk prices will result in about a $10.00 decrease in net returns to BST. A 2 to 4 percent decrease in response rate will result in a decrease of $24 to $48 in net returns. Thus, if a cow milking 15,000 lb returns $63.82 in additional net revenues when milk prices are at $11.00/cwt, then a $0.50 fall in milk prices, and a shortfall in the response rate by 2 to 4 percent would result in a $30- to $60-decrease in expected additional revenues. That is, expected returns to BST would drop from $63.82 to $3.82.

Criteria for Evaluation

In general, a positive net return may be sufficient to induce producers to adopt BST. If net revenues are negative, a loss is incurred and BST is not worth adopting. Using BST requires increased management, additional labor, and, initially at least, increased risk.

TABLE 6.1. Elasticity of the Change in Net Revenues with Respect to Change in Selected Variables.

	Percent
Response rate	1.82
Production base	1.97
Milk price	2.51
Cost of BST	-0.82
Feed costs	-0.55
Veterinary services and other costs	-0.15

TABLE 6.2. Guidelines to the Relative Sensitivity of Net Returns from BST Use due to Changes in Selected Parameters.

Parameter	*Unit Change*	*Approximate Change in Net Returns from BST Use*
Cost of BST	+ $0.01	- $ 2.00
Milk price	+ $0.01	+ $ 0.20
Milk price	+ $1.00	+ $20.00
Response rate	+ 1%	+ $12.00
Production base	+ 1,000 lb	+ $ 8.00
Change in feed costs	+ 1% per 10% increase	- $ 7.00
Veterinary services and other costs	+ $0.01	- $ 0.20

Producers may expect a positive return on an investment in BST, but how much is not clear. Some analysts have observed that farmers will require at least a 2:1 return on BST to adopt. If so, when BST costs $0.35 per cow per day, producers will require an increased net profit of at least $0.70 per cow per day. In the analysis above, net returns per cow per day are only $0.43 (i.e., $92.32/215 days). The cost of BST would have to be around $0.26 per cow per day to achieve, in the current example, a 2:1 return on BST.

An economic analyst might suggest that producers should pay up to the point where marginal costs are equal to marginal revenues. This would change the analysis substantially. By manipulation, with everything else being equal and with no returns to risk and management, producers could afford, theoretically, to pay as high as $0.77/cow/day for BST and still be better off. Of course, this type of marginal analysis should be applied to the entire farm operation--not just to BST for the dairy operation.

Other bottom line measures that could be used to evaluate the feasibility of using BST are net returns per cow per day, net returns over costs, and net returns to BST.

Net returns per cow per day = NR/ND,

where	NR	=	net revenues or net returns, and
	ND	=	number of days administered BST.

Net returns over cost = $NR/\Delta TC$,

where	NR	=	net returns, and
	ΔTC	=	change in total costs.

Net returns to BST = NR/CBST,

where	NR	=	net returns, and
	CBST	=	cost of BST.

From the original example above these ratios are as follows:

Net returns per cow	$92.32
Net returns per cow per day	$ 0.43
Net returns over costs	66.17%
Net returns to BST	1.23

Some Words of Caution

Having reached the "bottom line" in the analysis, a few words of caution are necessary. As mentioned previously, there are still many unknowns with respect to the effects of BST on dairying operations and with respect to the parameters of the analysis.

Response Rates. It is important to understand that published claims that BST increases milk production by 10 percent, 15 percent, or 25 percent usually refer to *average daily response rates during administration of BST.* If, for example, a producer administered BST for just one month during, say, midlactation, then the expected increase in milk production (e.g., 10%, 15%, 25%) would occur only during the one month of administration. If, during the particular month selected for BST administration, the cow gives, say 15 percent of the total milk for the lactation, then the response rate of BST for the *entire lactation* will be 15 percent of the claimed average daily response rate. For example, if the expected (or claimed) average daily response rate of BST is, say,

20 percent, and the cow is expected to give 15 percent of the total milk for the lactation during the time chosen for administration of BST, then the response rate of BST for the *entire lactation* will be 3 percent (0.15 x 0.20 = 0.03).

The distinction between average daily response rates and response rates for the entire lactation has been a constant source of confusion for many people. For example, when the results of the first trials of BST at Cornell University where published, it was reported in the popular press that BST could increase milk production by 25 to 40 percent. What was not reported was: 1) that these results were *average daily* response rates, and, 2) that it was only feasible to administer BST during mid- and late lactation. Apparently the Cornell trials had led to the conclusion that BST administered during early lactation was not as effective, and could possibly cause problems with conception (milking cows come into heat during early lactation and it is desirable to get them pregnant at that time).

A year in the life of a milking cow can conveniently be split into 4 production periods. A cow will produce different amounts of milk in the different production periods. The percentage of total milk produced in each period is a known and consistent result, as follows:

Production Period	*Average Days*	*Percent of Total Milk Produced*
Early	91	36
Mid	120	42.9
Late	95	21.1
Dry	59	0
	365	100

Since trial results indicate that it is feasible to administer BST only during mid- and late lactation, increases in milk production can occur only during the period when 64 percent of total milk is produced. Hence, if trial results indicate that *average daily* response rates to BST are 15 percent, then the response rate for the entire lactation (if administered throughout the mid- and late-lactation periods) will be 9.6 percent (0.64 X 0.15 = 0.096) as shown in the following tabulations:

	Response Rate (%) for Entire Lactation When BST Is Used:		
Average Daily Response Rate (%)	*Midlactation Only*	*Late Lactation Only*	*Mid- and Late Lactation*
5	2.15	1.06	3.2
10	4.29	2.11	6.4
15	6.44	3.17	9.6
20	8.58	4.22	12.8
25	10.73	5.28	16.0
30	12.87	6.33	19.2

In order to arrive at a response rate for the *entire lactation* of 12 percent, (and assuming that BST is administered for the 215 days of mid- and late lactation) average daily response rates would have to be at least 18.75 percent (12/0.64 = 18.75).

In the original example, a response rate based on a percentage increase in milk production was used. Some experimental results have shown that the response rate may be an absolute increase of about 8 lb per cow per day, regardless of the production base of the cow (Fallert et al. 1987). This would greatly affect the analysis in the example above. For example, if the response rate was 8 lb of milk per cow per day, then the increase in milk would be 1,720 lb during the period of BST administration (215 days at 8 lb/day), representing a 10.2 percent response rate for a cow milking 16,800 lb. Clearly, the percentage response rate would change for each cow with a different production base, i.e., 8 lb/day is equivalent to a 14.3 percent response rate for a cow milking 12,000 lb and 8.6 percent for a cow milking 20,000 lb. Figures 6.4, 6.5, and 6.6 show the change in net revenues for various response rates, milk prices, and cost of BST associated with nonproportional response rates.

Two things should be noted from these figures. First, the range of changes in net revenues is much smaller for nonproportional response rates than for proportional response rates. Second, the rate of change between production bases is much less dramatic for nonproportional response rates than for proportional response rates. Therefore, a non-proportional response rate of 8, 10, or 12 lb/cow would mean that BST was feasible--or not feasible--for a much larger percentage of the herd.

Some scientists feel that the response rate to BST may be a mixture of proportional and nonproportional rates. The response rate could be, for example, 700 + (0.06 x PB). For a cow milking 16,800 lb, this would be an increase in production of 1,708 lb, a 10.16 percent response rate. Again, response rates will differ for each production base, but the feasibility analysis will not be nearly as sensitive to changes in the production base as was the case with proportional response rates.

Apart from the uncertainty associated with measuring the response rate in terms of proportional or nonproportional increases in milk production, there is also considerable uncertainty associated with the

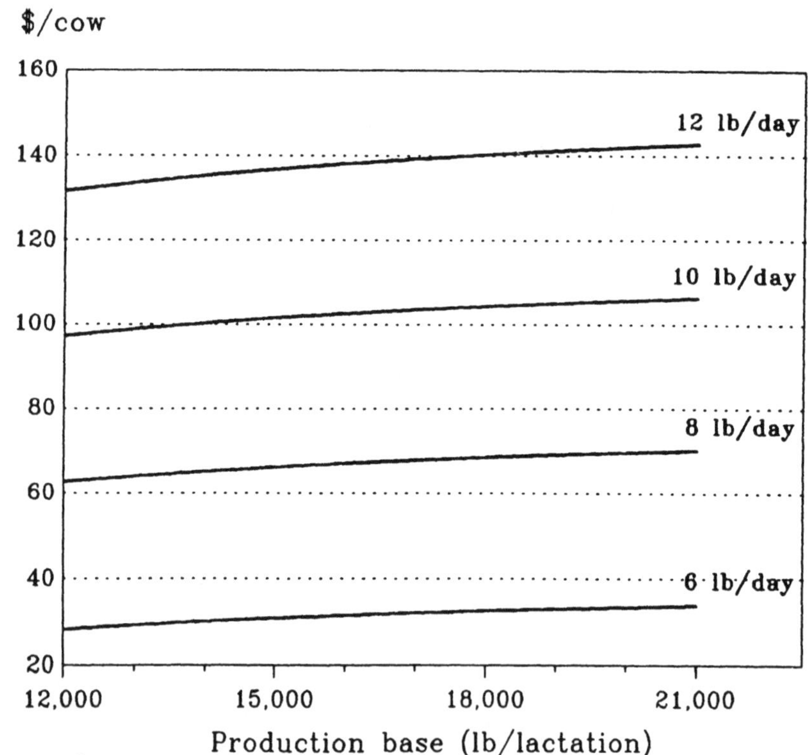

Assumptions:
Milk price: $11.50/cwt
Change in feed costs: 6% per 10% increase in milk production
BST cost: $0.35/cow/day
Additional costs: veterinary, labor, milk-transport, etc: $0.67/additional cwt produced

FIGURE 6.4. Change in net revenue ($/cow/lactation), assuming BST use for 215 days, by four response rates (6 to 12 lb/day) and four production bases (12,000 to 21,000 lb/lactation).

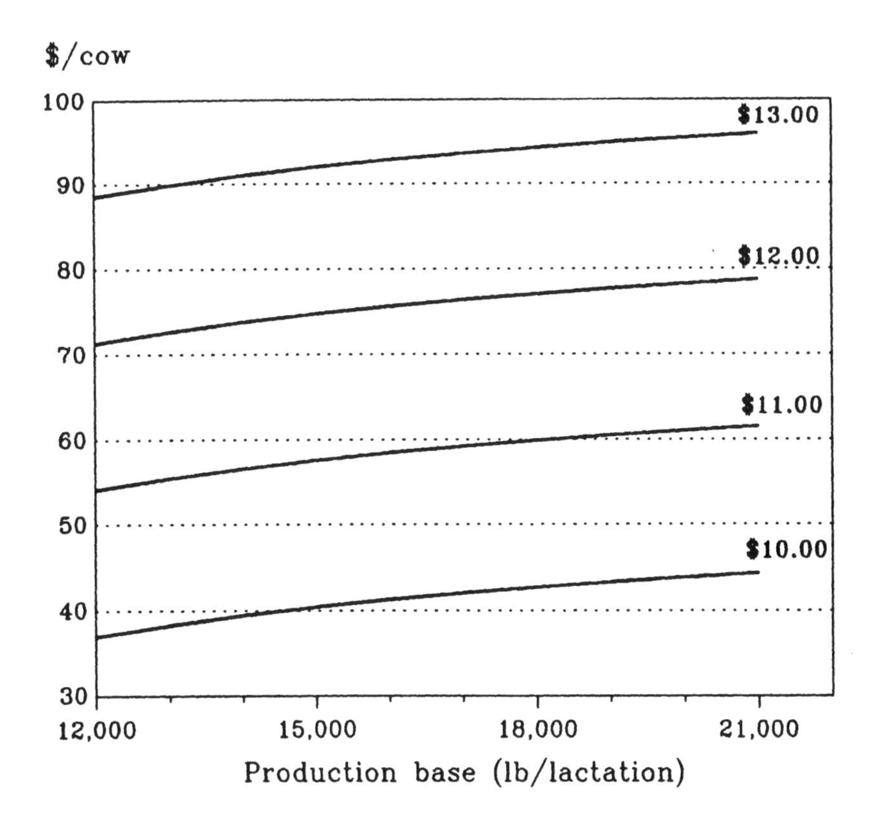

Assumptions:
Change in feed costs: 6% per 10% increase in milk production
BST cost: $0.35/cow/day
Additional costs: veterinary, labor, milk-transport, etc:
$0.67/additional cwt produced

FIGURE 6.5. Change in net revenue ($/cow/lactation), assuming BST use for 215 days, with a response rate of 8 lbs/day by four milk-price levels ($10.00 to 13.00/cwt) and four production bases (12,000 to 21,000 lb/lactation).

actual response rates themselves. Aggregated trial results from trials carried out over the last 10 years suggest that the range of responses to BST can vary anywhere between 0 and 30 percent. However, because individual trials are rarely reported in academic journals, it is almost impossible to estimate the level of variability (risk) associated with this broad range of responses.

Interpretation of published data from nine long-term trials is currently debated in the magazine *Large Animal Veterinarian.* Scientists from Monsanto state that the appropriate representative range of response in BST-treated herds is 10 to 25 percent. Virginia Tech veterinarian Dr. David Kronfeld (1987) disagrees, estimating from the same data that response from 95 percent of the herds will range from minus 2.3 percent to 29.4 percent. Mean response, according to Kronfeld, will be 13.5 percent and the standard deviation 7.6.

From these values, Kronfeld predicts that with a breakeven point of 11.4 percent, 39 percent of all herds on BST treatment will lose money, reinforcing the opinion that considerable monetary risk may be involved with BST use.

The Monsanto scientists argue that standard error, rather than standard deviation, should be used in measuring risk. They wrote: "the standard deviation that Kronfeld uses is a measure of the variability in milk production of individual cows, not the mean response."

Standard error (SE) and standard deviation (SD) are related, dependent upon sample size (N), with standard error being the standard deviation divided by the square root of the sample size, i.e.,

$$SE = SD/\sqrt{N}$$

where

SE = standard error
SD = standard deviation
N = sample size

The standard deviation for this data is 7.6, so Kronfeld's estimate of risk will hold true for relatively small herds, but will be too small for larger herds. If the breakeven response rate is 12 percent, the proportion of herds for which BST would not be worthwhile ranges from about 40 percent for herds less than 20 cows down to 30 percent for herds of 80 to 100 cows. However, if the breakeven response rate is 10 percent, these estimates are much lower. The level of risk associated with using BST is very sensitive to both the value of the mean response rate and the breakeven response rate. In general, the larger the difference between the mean response rate and the breakeven point, the smaller the risk associated with using BST.

In addition to the uncertainty associated with the response rate of cows on BST, there is some evidence that first-calf heifers respond less (or differently) to treatments of BST than do older cows. Finally, BST response rates may not be the same in the field as they are under

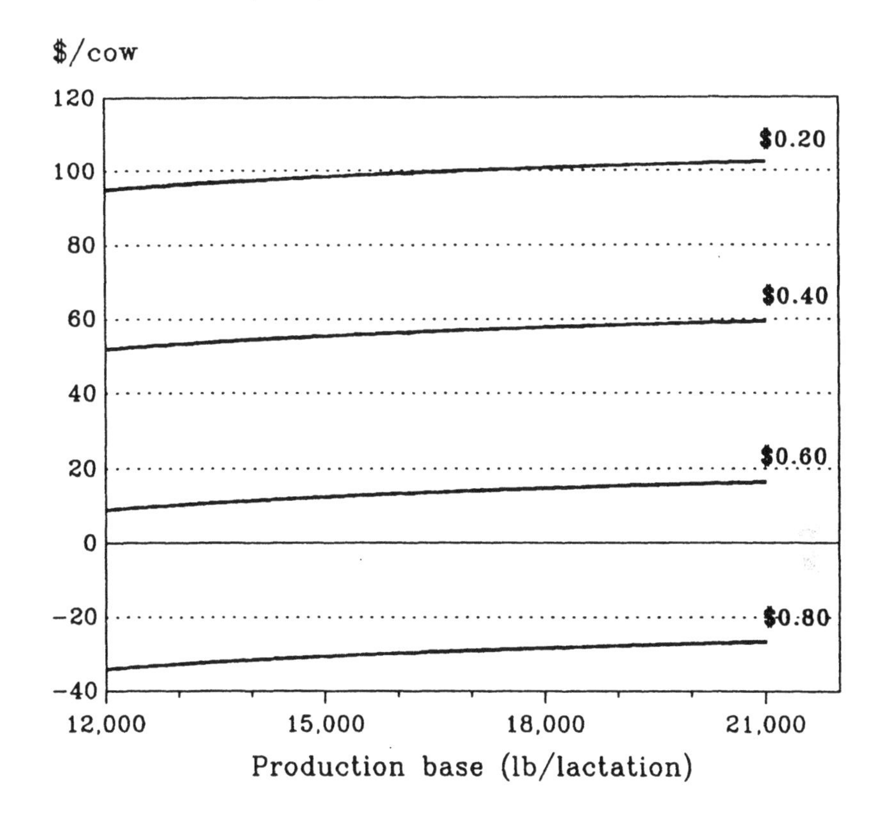

Assumptions:
Milk price: $11.50/cwt
Change in feed costs: 6% per 10% increase in milk production
Additional costs: veterinary, labor, milk-transport, etc:
$0.67/additional cwt produced

FIGURE 6.6. Change in net revenue ($/cow/lactation), assuming BST use for 215 days, at a response rate of 8 lb/day, by four levels of BST cost ($0.20 to 0.80/cow/day) and four production bases (12,000 to 21,000 lb/lactation).

carefully managed research conditions. As a worst-case scenario, it may be prudent to assume that BST response rates could be as much as 25 percent less than research results (Fallert et al. 1987).

Effects on Feed Consumption

The assumption that the increase in feed costs is about 6 percent per each 10 percent increase in milk production is based on least-cost feed mix rations of the mid-1980s, and on general feed-cost increase data. Obviously, changes in the price of feed will change both the base feed costs and the calculated feed-cost increases. A prudent analysis would include a worst-case scenario in these calculations to allow for unexpected fluctuations in feed costs due to drought, policy changes, etc.

Cost of BST

The companies producing BST have not been forthcoming in predicting the likely cost (price) of BST. Most indicate somewhere between 25 and 75 cents/cow/day. Time-release injections or implants, researched by some companies, appear to be more convenient and are likely to produce results similar to daily injections. But they are also likely to be more expensive. Since the economic feasibility of using BST is very sensitive to the cost of BST, a worst-case scenario would include an analysis of the maximum amount that a producer would be prepared to pay for BST.

Required Rate of Return to Management and Risk

It has been indicated that, in general, a positive net return on BST use may be sufficient to conclude that BST use is feasible. In many herds, substantial changes in management will be required to use BST, and with the element of risk due to uncertainty of response rates, etc., it may be prudent to include in the analysis a return to management and risk sufficient to ensure that large losses do not occur if BST proves to be nonviable for a particular operation. As a rule of thumb, one could include a 2:1 return on BST in the worst-case scenario. That is, one could expect a return of at least twice the cost of the BST. An approximation of this would be to double the cost of BST in the calculation.

Evaluation of BST: Measuring Response Rates

Apart from carrying out an analysis of the cost effectiveness of BST or on the feasibility of its use on some cows in a dairy operation, producers should give some thought to how they will evaluate its effects. In order to evaluate response rates, a milk-o-meter or other device that weighs the increase in milk production may be necessary. In addition, it will be necessary to monitor increased feed intake by the BST-treated cows in order to evaluate increased feed costs. If producers are unable to measure the actual effectiveness of BST, then adopting BST may not be feasible for that reason alone. Producers will not be able to optimize its use without such an evaluation.

Changes in Other Costs

Producers should be careful to evaluate other costs associated with administering BST. There may be some additional labor costs, although many producers will be able to utilize current labor. Because of the additional stress put on cows producing more milk, additional veterinary costs may be necessary. These should be carefully monitored. Increased difficulties in settling some cows treated with BST should be noted and the associated costs included in the feasibility analysis. Of course, much of this will be unknown until one or two lactations of treatment have occurred. From that perspective, it may be prudent to include some "fudge" costs in the analysis and to experiment with BST on a small proportion of the herd at first.

Finally, it should be emphasized that the basic evaluation explained here is just that, a basic evaluation. Nothing can substitute for a full budgeting exercise that puts using BST use in the perspective of a whole-farm operation.

BST, Assessments/Quotas, and Farm-management Decisions

The controversy surrounding BST has stimulated debate in many states, and some states have actually enacted legislation intended to limit its use when it becomes commercially available. The U.S. Congress is unlikely to attempt to control or limit the use of BST unless there is some human health or environmental reason to do so. However, Congress *is* concerned about the potential for excessive surpluses of milk that may be produced if BST is widely adopted and diffused when it becomes commercially available. Hence, the 1990 Farm Bill provides for substantial control over excess production. The possibilities are as follows:

> *Assessments on All Milk Produced in Excess of 7 Billion Pounds.* Under the 1990 Farm Bill, USDA is required to estimate, by November 20 of each year, the upcoming year's surpluses. Beginning in 1991, with the November 20 announcement, USDA will raise or lower milk support prices accordingly and calculate the amount producers will be assessed for surpluses in excess of 7 billion pounds. (These assessments would be added onto the already mandated assessments of $0.1125/cwt for 1992-1995.) Refunds would be provided to all producers who can show that they have not increased production from the previous year.
>
> *Class IV Pricing as an Alternative Classification of Milk.* Presumably, this type of program will be accompanied by some sort of quota on all producers. Milk in excess of an individual's quota would be classified as Class IV milk, and would be priced at some percentage of the support price.
>
> *Target Prices/Deficiency Payments.* It can be assumed that this type of program will replace the current government purchase program, thus phasing out CCC purchases. Under a price-deficiency payment program, target prices would be set at about the same level as current support prices. However, milk would be sold at retail at a specified market-clearing price. The difference between the lower retail market-clearing price and the target price would be paid by the government.
>
> Presumably this program would also be accompanied by either a maximum quantity of milk or a quota. Milk products exceeding the maximum or the quota would not be eligible for the deficiency payment, and will command a lower price.

These programs will affect the feasibility of using BST on the dairy farm. Each producer's decision must consider the type of program implemented, the magnitude of the assessment, the price differential, and the costs of production.

Evaluating Assessments and Quotas

The first decision that producers must face is the question of whether or not it is worth producing surplus or over-quota milk, given that this milk will receive a lower price--either in the form of an assessment or lower-class price.

In order to evaluate the effect of quotas or assessments on profits, one must estimate the change in net returns associated with increased production, in much the same way that one estimates the feasibility of using BST. (There is one essential difference, however. A negative change in net revenues or profits as a result of expanded production is not necessarily an indication that expanded production is not worthwhile. A producer may, for example, figure that it is worth taking a loss in the short term in return for a longer-term gain.)

The change in net revenues associated with expanding production, when the additional production receives a lower price or when an assessment is levied on all milk produced when it exceeds its historical base, is given by

$$(1)\quad \Delta NR = NR^E - NR^O$$

where NR^E is the net revenues associated with expanded production and NR^O is net revenues associated with nonexpansion.

Assessments on Milk Production. Under the current program of assessments on milk production, all milk is assessed a fixed fee/hundredweight. The assessments are refunded if a producer can show that milk production on his/her farm did not exceed that of the previous year. Hence, the decision a producer faces is if it is worth while maintaining current production levels or expanding production.

One way of evaluating the decision is to determine if there will be an increase in net revenues (profits) from the increased milk production. This can be assessed as follows:

Let $NR^E = Q(P\text{-}A\text{-}C) + \Delta Q\,(P\text{-}A\text{-}MC)$, and
$NR^O = Q\,(P\text{-}C)$,

where NR^E = net revenues from expanded production,
NR^O = net revenues from nonexpansion,
Q = base quantity of milk produced,

P = prevailing price of milk,
A = assessment,
C = average cost of producing a hundredweight of milk, and
MC = marginal cost of production.

Substituting these two expressions into equation (1) gives:

$$(2) \quad \Delta NR = Q\,(P\text{-}A\text{-}C) + \Delta Q\,(P\text{-}A\text{-}MC) - Q\,(P\text{-}C)$$

$$= \Delta Q\,(P\text{-}A\text{-}MC) - QA.$$

If $\Delta NR = 0$ (break-even) then

$$\Delta Q\,(P\text{-}A\text{-}MC) = QA, \text{ and}$$

$$(3) \quad \Delta Q/Q = A/(P\text{-}A\text{-}MC).$$

Thus, if ΔNR is to be positive (increase in net revenues) then production must increase by $\Delta Q/Q$ to make it worth while. In other words if $\Delta Q/Q < A/(P\text{-}A\text{-}MC)$, then there will be a net loss in revenues due to expansion in milk production. For example, if current assessments are \$0.05/cwt, price is \$10.50/cwt, and MC is \$6/cwt then $\Delta Q/Q = 1.12$ percent. Therefore, production must increase at least 1.12 percent for expansion to be worth while. If assessments increase to \$0.1125/cwt, then $\Delta Q/Q = 2.56$ percent. Figure 6.7 plots the breakeven increases in production for various assessments and profit (net revenues) per hundredweight levels.

Quotas and Two-tier Pricing. When quotas are imposed on a base level of production and the price received for over-quota milk is lower than the current market or prevailing price, the evaluation changes only slightly.

Let $NR^E = Q(QP\text{-}C) + \Delta Q\,(OQP\text{-}MC)$ and
$NR^O = Q(QP\text{-}C)$

where QP = quota price, and
OQP = over-quota price.

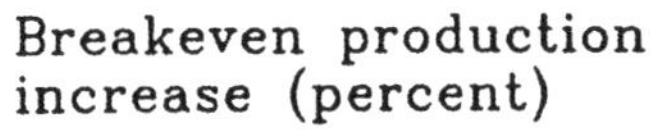

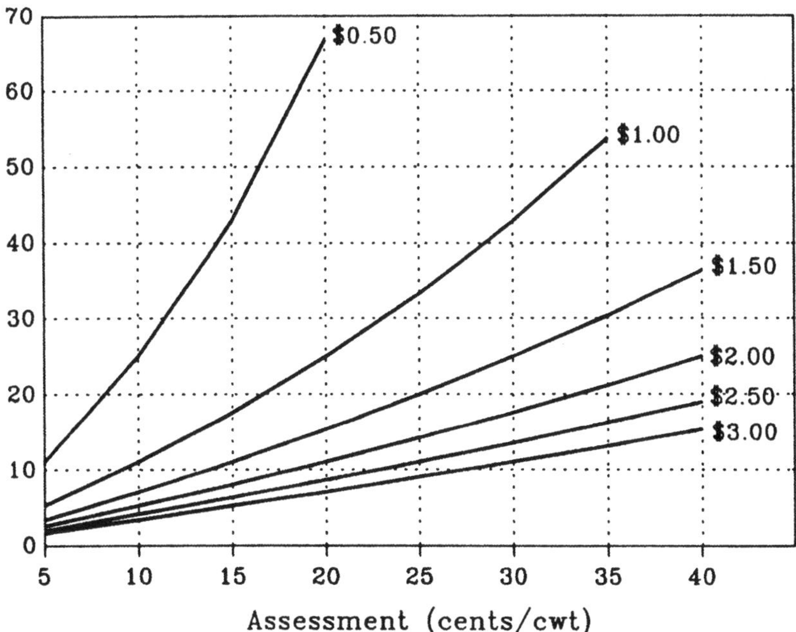

FIGURE 6.7. Minimum (breakeven) increases in production (%) required to achieve six profit margins ($0.50 to 3.00/cwt) and eight assessment levels (5 to 40 cents/cwt).

Substituting into equation (1) gives:

$$\Delta NR = Q(QP\text{-}C) + \Delta Q(OQP\text{-}MC) - Q(QP\text{-}C)$$
$$= \Delta Q(OQP\text{-}MC).$$

If $\Delta NR = 0$ (breakeven) then

$$\Delta Q(OQP\text{-}MC) = 0, \text{ and}$$
$$OQP = MC.$$

Under this scenario, the break-even point is given by equating over-quota price to the marginal cost of production. In other words, as long as over-quota price exceeds the marginal cost of production, then the change in net revenues will be positive, and expansion will be worthwhile.

Size and Scale Adjustments. If producers, using the methods described in the previous section, can ascertain that it is worth continuing to produce over-quota or surplus milk then they will, in general, be able to profitably use BST on those cows for which it is determined that it is feasible.

However, if it is determined that a loss in net revenues occurs [either $\Delta Q/Q < A/(P\text{-}A\text{-}C)$ or $OQP < MC$] the producer must decide whether it is worth taking a short-term loss in return for a longer term gain. But this *does not mean* that these producers will be unable to use BST.

It is important to bear in mind that while BST increases milk production, it *also increases* production efficiencies. BST-supplemented cows do require more feed in order to produce the additional milk, but the additional feed required is proportionately less relative to the extra milk produced. Put in another way, the extra milk produced by a BST-supplemented cow is worth more (at current prices) than the extra feed needed to produce it.

Many studies have shown that feed efficiencies with BST use are increased 6 to 17 percent, while the corresponding increase in feed costs are only 5 to 8 percent (Chilliard 1988). This means an effective decrease in feed costs per hundredweight of milk produced, and therefore a lower average cost of production.

The following results show that a producer can feasibly reduce the number of cows in the herd, produce *less* milk, and *increase* net revenues. The number of cows supplemented with BST required to produce the same revenue as a herd *not* supplemented with BST is given by

$$M = N[APB(MP\text{-}AC)]/APB^T(MP\text{-}AC^T)]$$

where

- M = number of cows supplemented with BST,
- APB = average production base of herd not using BST,
- APB^T = average production base of herd supplemented with BST,
- MP = milk price,
- AC = average cost of production of herd not using BST,
- AC^T = average cost of production of herd supplemented with BST, and
- N = number of cows in herd.

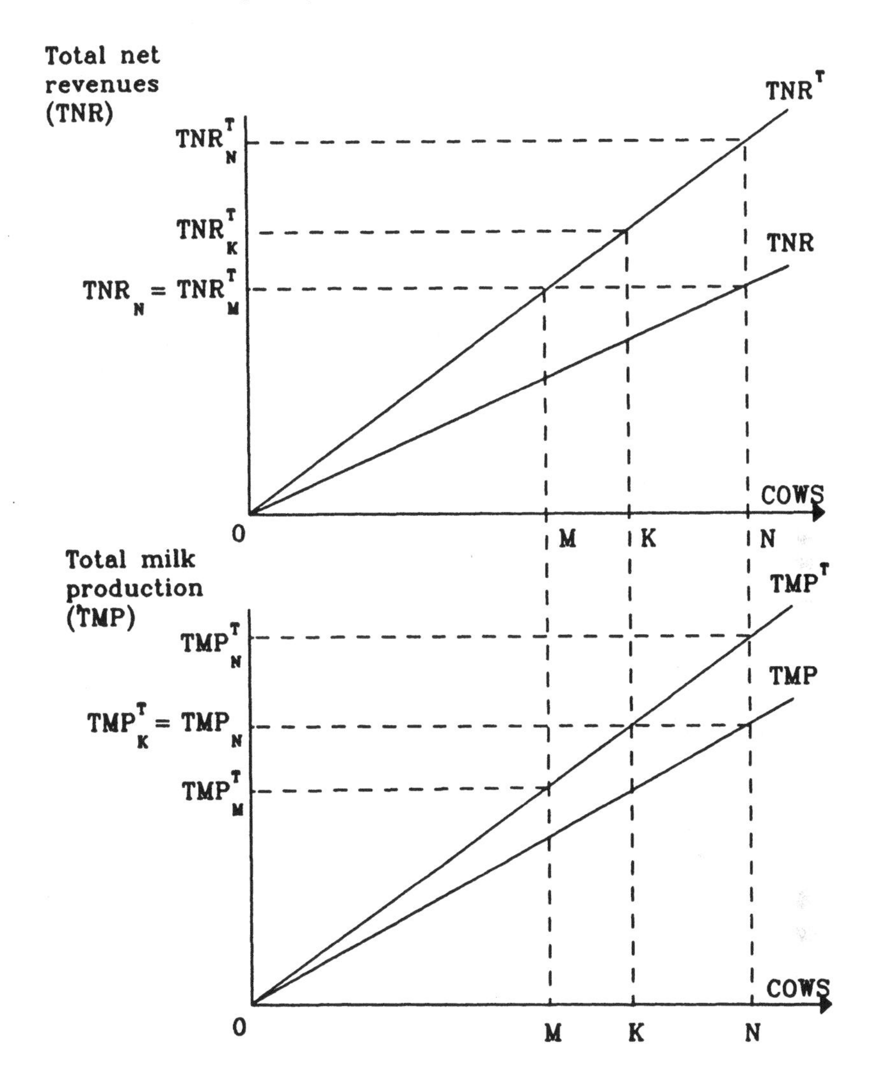

FIGURE 6.8. Total net revenues and total milk production vs. number of cows, with and without BST.

The number of cows supplemented with BST required to produce the same volume of milk as a herd not supplemented with BST is given by

$$K = N[APB/APB^T]$$

Proofs of these formulas are given in Appendix 6.A, and the results diagrammed in Figure 6.8. In Figure 6.8, there are N cows in the herd not supplemented with BST. TMP_N is the amount of milk produced without BST, and TNR_N is the total net revenue achievable. If *all* N cows are supplemented with BST, then total milk production can be increased to TMP_N^T, and total net revenues increased to TNR_N^T.

It requires only M cows supplemented with BST to achieve the same total net revenues as a herd of N cows not supplemented with BST, and they will produce TMP_M^T pounds of milk, which is less than TMP_N.

It requires K cows supplemented with BST to achieve the same total milk production as a herd of N cows not supplemented with BST, and net revenues will increase to TNR_K^T, which is more than TNR_N.

Thus, under a quota system, a producer would be better off financially to use BST and reduce the size of the herd, than to *not* use BST and keep the same number of cows.

Summary

The essence of an evaluation of BST for onfarm use is to examine the consequences on net revenue. Ideally a whole-farm analysis should be conducted, but much insight can be gained from a partial-budget analysis focused on the dairy enterprise alone. Several rules-of-thumb for examining the sensitivity of net returns from BST use due to changes in selected parameters of relevance to the analysis are derived--e.g., milk prices, feed costs, cost of BST, response rate to BST, etc. Some precautions are also given for anticipating some of these parameters. Finally, this chapter discusses how the basic analysis can be used to evaluate alternative production decisions under selected policy scenarios, such as producer-assessment schemes and quota programs.

Acknowledgments

I appreciate the helpful comments of Dr. W. McSweeny on an earlier draft of this chapter.

References

Fallert, R. F., T. McGuckin, C. Betts, and G. Bruner. 1987. *bST and the Dairy Industry: A National, Regional and Farm-level Analysis*. Report #579. Washington, D.C: Economic Research Service, USDA.

Chilliard, Y. 1988. "Long term effects of recombinant bovine somatotropin (rBST) on dairy cow performance: A review." Pp 61-87 in K. Sjersen, M. Vestergaard and N. Sorenson (eds). *Use of Somatotropin in Livestock Production*. London, UK: Elsevier Applied Science.

Kronfeld, D. S. 1987. "The challenge of BST." *Large Animal Veterinarian* 42(8)14:17.

Chapter 6 Appendix

The net revenue earned per cow is:

Equation 1 $NR_i = GR_i - TC_i$

where NR_i = net revenue per cow,
GR_i = gross revenue per cow, and
TC_i = total cost of production per cow.

Gross revenues per cow are given by:

Equation 2 $GR_i = P(PB_i)$

where PB_i = production base in pounds of milk per cow, and
P = price received for milk.

Total costs are given by:

Equation 3 $TC_i = C_i(PB_i)$

where C_i = cost of milk production per pound.

Substituting Equation 2 and Equation 3 into Equation 1 gives:

Equation 4
$$NR_i = P(PB_i) - Ci(PB_i)$$
$$= PB_i(P - C_i)$$

Total net revenue of the herd of N cows:

Equation 5
$$TNR = \sum_{i=1}^{n} PB_i (P-C_i)$$

or

Equation 6
$$TNR = N[APB (P-AC)]$$

where TNR = total net revenue for the herd,
N = number of cows,
APB = average production base (lb/cow), and
AC = average cost of milk production.

For a herd in which BST is not used, denote net revenue per cow as

Equation 7 $NR_i = PB_i(P\text{-}C_i)$

where NR_i, PB_i, C_i denote net revenue, production base, and cost of production without supplemental BST.

Total net revenues for the herd will be:

Equation 8 $$TNR = \sum_{i=1}^{n} PB_i\,(P\text{-}C_i)$$

or

Equation 9 $TNR = N\,[APB\,(P\text{-}AC)].$

For a herd in which M cows are supplemented with BST, denote net revenues as:

Equation 10 $NR_i^T = PB_i^T(P\text{-}C_i^T)$

where $PB_i^T > PB_i$ and $C_i^T < C_i$.

Therefore $NR_i^T > NR_i$

Total net revenues for a herd of size M cows supplemented with BST is:

Equation 11 $$TNR^T = \sum_{i=1}^{M} PB_i^T\,(P\text{-}C_i^T)$$

or

Equation 12 $TNR^T = M\,[APB^T\,(P\text{-}AC^T)].$

If the two herds are the same size (i.e., M = N) then

$$TNR^T > TNR.$$

In order to achieve the same total net revenue in each herd, set Equation 9 equal to Equation 12

i.e., $$TNR = TNR^T$$

or upon substituting,

$$N\,[APB\,(P\text{-}AC)] = M\,[APB^T\,(P\text{-}AC^T)].$$

Therefore, the number of cows supplemented with BST required to produce the same revenue as a herd not using BST is:

Equation 13 $$M = \frac{APB\,(P\text{-}AC)}{APB^T\,(P\text{-}AC^T)} \bullet N.$$

Since $APB < APB^T$ and $AC > AC^T$, then $APB\,(P\text{-}AC) < APB^T\,(PAC^T)$, and therefore $M < N$. Total milk production in the herd of N cows not being supplemented with BST is given by:

Equation 14 $$TMP = N\,(APB).$$

Similarly, total milk production for a herd of M cows supplemented with BST is:

Equation 15 $$TMP^T = M\,(APB^T).$$

Substituting for M in Equation 15 from Equation 13 gives:

$$TMP^T = \frac{APB\,(P\text{-}AC)}{APB^T\,(P\text{-}AC^T)} \bullet N \bullet APB^T.$$

Equation 16 $$= \frac{N \bullet APB\,(P\text{-}AC)}{P\text{-}AC^T}.$$

Since $(P\text{-}AC^T) > (P\text{-}AC)$, then $(P\text{-}AC)/(P\text{-}AC^T) < 1$, and therefore $TMP^T < TMP$.

Hence, a smaller herd of M cows supplemented with BST will produce the same amount of revenue and less milk than a herd of N cows not supplemented with BST.

These results suggest that if BST can be used to reduce the size of the herd, and produce *less* milk for the same net revenues, then BST can also be used to reduce cow numbers to a level that will produce the same amount of milk and *increase* revenues.

Setting total milk production of the two herds equal (i.e., setting Equation 14 and Equation 15 equal) gives:

$$TMP = TMP^T, \text{ and}$$

$$N \bullet APB = K \bullet APB^T$$

where $K \neq M$.

Thus,

Equation 17 $$K = \frac{APB}{APB^T} \bullet N.$$

Since $APB^T > APB$, then $K < N$.

In addition, from (13), since $(P\text{-}AC) < (P\text{-}AC^T)$, then $K > M$.

Therefore, $M < K < N$.

7

Potential Adoption and Diffusion of BST Among Dairy Farmers

Robert D. Yonkers

Diffusion of technology is the last step of technological change. The process begins with basic research of underlying scientific principles, and moves through the invention stage as applied researchers develop basic concepts into technical applications. Actual technical innovation comes with commercial development and introduction of such applications. The extent and speed with which a new technology is adopted among individual firms describes diffusion.

The need to provide information about potential impacts of BST has led to research which estimates the rate of adoption among dairy farms, the focus of this chapter. The first section discusses factors affecting technology adoption and the role of technological change in the dairy industry over the years. Then, methods used to research adoption and diffusion will be discussed, followed by a survey of BST-adoption research to date.

Factors Affecting the Rate of Technology Adoption

Technology can cause old products to be replaced by new ones, create new inputs to improve old ones, or otherwise affect the production process. Technological change enables a firm to produce either greater output and/or product quality with the same level of inputs, or the same output and/or product quality with less of at least one input (Just, Schmitz, and Zilberman 1979). The result is either lower costs of production per unit of output or increased value (quality) of that unit of output.

Farm managers operate in a competitive industry; prices received for their farm outputs are set by the market. Therefore, they can benefit from adoption of new technology that lowers costs. However, there are factors other than economics that may influence an individual farm manager's decision to adopt technology. Rogers (1983) suggests the following criteria be added to any relative economic advantage of one technology over another:

1. Its compatibility with existing resources and current production and management practices.
2. Its complexity, either in use or in understanding how it works.
3. Its cost, particularly capital investments (including human resource development).
4. Its adaptability on a small scale, permitting limited commitment until its value for full-farm adoption is established.
5. The ease with which information about its function, benefits, and risks can be communicated to potential users.

Any or all of these factors may affect an individual dairy farm manager's decision whether or not to adopt BST. Each farm manager has an objective function, which may include philosophies about farming and farm life, and attitudes toward technological change. How a particular technology's characteristics affects that function is known only to that manager. Economic analysis of new technology relies on assumptions regarding future prices and quantities. Such key variables are not known with certainty beforehand, and may lead to decisions based on risk preferences. Research trials using new technology on a limited basis may provide information for a farm manager's decision to adopt. Lack of access to information about innovations, including costs and benefits, has been cited as a source of failure to adopt (Swanson et al. 1986).

Rates of technological change vary across sectors of the economy. Such variation can be related to industry structure, availability of information, and historical patterns of technology adoption. Emerging technologies with applications for the production of milk should be viewed in the context of past technological change in the dairy industry.

Technological Change and the Dairy Industry

Milk production, once nearly universal on all farms in the United States, has become a specialized form of commercial farming. A number of technological advances contributed to this evolution, most of which were developed exogenously to the dairy industry. Two significant advances were development of improved rural transportation systems and rural electrification. The former facilitated bulk marketing of raw milk off the farm and the latter provided the opportunity to utilize power equipment for milking and care of dairy livestock. In addition, access to formal education and technical training has improved the abilities of dairy farm managers, especially their understanding of the use of technology, to increase productivity.

The history of the dairy industry in terms of onfarm production and management practices may be characterized by technological change. The early use of information technologies [e.g., dairy cow rations balanced to meet nutritional requirements, and production records provided through Dairy Herd Improvement (DHI) associations] allowed farm managers to base production decisions on more objective analysis. Mechanical milking machines are found on nearly every dairy farm today. The use of artificial insemination (AI) technology has increased on dairy farms, freeing resources once used to house and care for bulls.

Such technologies have allowed labor to focus on milk harvest; milk productivity per man-year was in 1983 seven times that of the 1930s (Manchester 1983, p 49). Productivity, measured as annual milk output per cow, has also grown due to technological change. The rate of genetic progress has increased due to the widespread use of AI and DHI records and the limited use of embryo transfer. Improvements in feed-harvesting, -handling, and -marketing systems and the ease of testing feedstuffs for nutritional value has led to higher-quality rations for dairy animals, promoting faster rates of growth and increased feed efficiency for milk production. Computerized record-keeping systems adapted to dairy farms provide key information for management decisions on herd health, feeding, breeding, and culling.

Although the dairy industry has a history of adopting technology, diffusion of new technologies is not universal among dairy farms. This has been due largely to those characteristic factors of each technology that affect an individual farm manager's decision to adopt. Table 7.1

TABLE 7.1. Date of Introduction and Degree of Adoption, by 1985, of Selected Technologies on Dairy Farms in the United States.

Technology	*Year of Introduction*	*Use in 1985 (%)*
Scientific feeding	1920	70.0
DHI testing	1925	45.0
Mechanical milking	1930	100.0
Artificial Insemination	1940	70.0[a]
Electronic farm accounting	1960	30.0
Milking parlors	1960	25.0
Bulk tanks	1960	100.0
Freestall housing	1960	25.0
Embryo transfer	1980	0.5

[a] Percent of all dairy cows and heifers bred artificially.

Source: Conneman 1987.

details the year of introduction and level of diffusion by 1985 for many of the dairy farm technologies discussed above.

To observe and report adoption rates and level of diffusion at a point in time after commercial introduction of an innovation is a straight-forward procedure. Predicting such information before actual commercial availability requires other research approaches.

Researching Technology Adoption

Research describing the potential impacts of technological change is increasingly in demand as technology becomes publicly debated and, therefore, is a public-policy issue. Rate of adoption is a key factor in technology-impact assessment and the most problematic, since the speed and extent to which a new technology is used will not be known with certainty until after the fact. Such is the case with agricultural applications of biotechnology in general and BST in particular.

Evaluating the impact of technological change by observing the speed and extent of adoption among farms after allowing a reasonable period for adoption is referred to as *ex post* analysis. Griliches (1957) was the first to quantify these observations with a logistic function as follows:

$$P = K/[1 + \exp(-a-bt)]$$

where
P = the level of diffusion,
K = the maximum level of diffusion,
a = a constant,
b = rate of acceptance, and
t = time.

Plotting this relationship between the level of diffusion and time elapsed since innovation availability yields a curve that can be described as S-shaped. Subsequent research found similar relationships for other agricultural technologies, although different estimated values for the parameters a and b and assumed value for K will produce variations in the shape of the curve (Figure 7.1).

This approach cannot be used for projecting potential impacts of emerging technologies prior to introduction and adoption. Such analysis requires *ex ante* assessment, where key variables have not yet been observed and are not known with certainty. *Ex ante* analysis applied to social science research in macroeconomic contexts is a relatively new methodology generated by requests for environmental- and technology-impact assessments by public institutions (Buttel and Geisler 1987). The functional form used by Griliches (1957) can estimate potential adoption and diffusion. However, such *ex ante* analysis requires assumed values for the key logistic function parameters a and b and K (Figure 7.1). This uncertainty about adoption rates makes potential-impact analysis problematic.

Impact assessment requires methodologies for obtaining data regarding potential adoption by farms and industry diffusion. Nowak (1987) describes four methods of obtaining data needed to estimate, prior to its introduction, the adoption and diffusion rate of a new technology:

1. Determine the behavioral intent of the potential adopter.
2. Examine farm characteristics, especially the distribution of constraints, related to technology adoption.

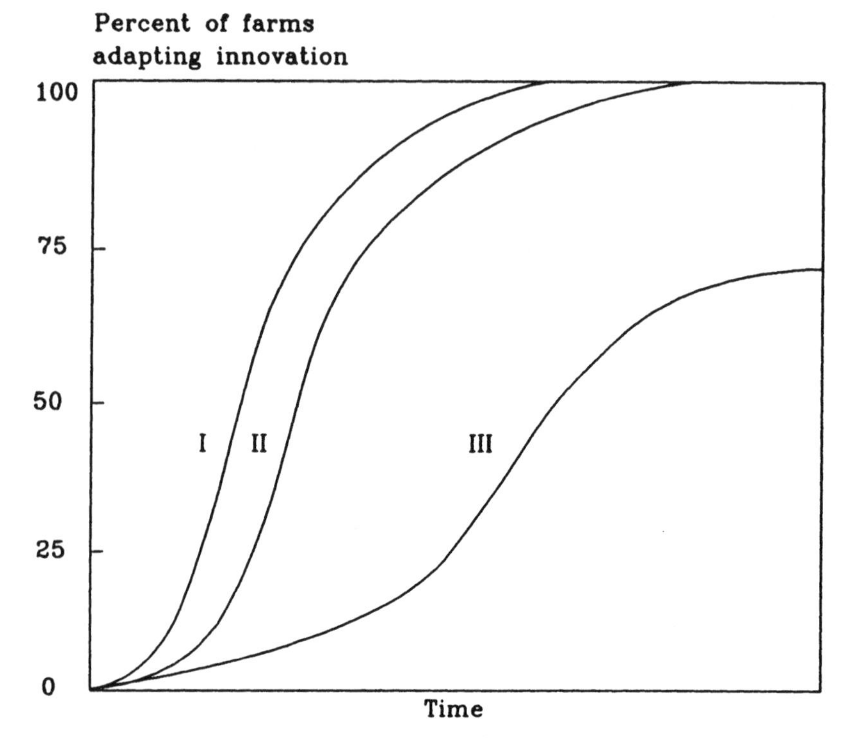

Source: Koch, p. 235. Adapted by permission of Prentice Hall, Englewood Cliffs, NJ.

FIGURE 7.1. Logistic Curves Representing Hypothetical Patterns of Adoption of a New Technology.

3. Look for patterns of behavioral consistency (e.g., has the farm adopted other technologies or management strategies similar in requirements for the technology under study?).
4. Seek expert opinion from scientists and industry analysts on the potential rates of adoption and diffusion.

A common feature of these methods is the need to use a survey to obtain data. No single method is clearly the most appropriate means to evaluate potential adoption and diffusion rates, and all have been used for BST-impact assessment.

Pre-1986 Studies on Potential Adoption of BST

The commercial potential of recombinant-BST did not become apparent until the early 1980s (Bauman et al. 1982). Bovine somatotropin remained largely a topic for the scientific literature until 1985. The few early agricultural press reports of BST were based on data from a limited number of small-scale research trials, and indicated that commercial availability was years away. As a result, detailed information about BST was not available to dairy farm managers. Surveys of potential BST adopters during this period required the presentation of information about the technology before questions about its use could be asked.

Kalter et al. (1984) conducted the first research of this type. A hypothetical Cooperative Extension "Fact Sheet" was developed, along with a fictitious BST advertisement in the style of a leading dairy farm publication. This information was presented to two samples of dairy farm managers. A random sample of 40 individuals was personally interviewed in seven representative New York dairy counties. A second, statewide random sample of 1,025 questionnaires was mailed; 133 dairy farm managers (13%) returned usable responses. For purpose of analysis, the two samples were combined, and, based on comparisons with secondary and other primary data, deemed representative of dairy farms in New York.

A majority, 61 percent, of the respondents indicated that they were at least somewhat favorably inclined to accept BST. Asked when they would adopt, 13 percent of all respondents reported they would never use BST; 49 percent indicated they would adopt the technology during the first year of commercial availability (another 2% gave no response; the remaining respondents indicated adoption between one and five years following availability). Price of BST was an important consideration among respondents. Nearly one-half (47%) said they would be less likely to adopt if the price of BST increased 32 percent, while a price decrease of 41 percent increased the likelihood of adoption for 40 percent of respondents.

In this New York study, analysis of variance was used to test for differences among early adopters (within one year), middle adopters (one to five years) and late adopters (after five years or never). Respondents indicating early adoption of BST tended to be younger than middle or late adopters (45.5 years of age, versus 49.1 and 48.0, respectively), but only at a 25 percent level of statistical significance. Level of output per cow was not related to speed of adoption, but herd size was. Early and middle adopters had much larger average herd sizes (72 and 70 cows, respectively) than did late adopters (49 cows).

The only other statistically-significant difference was that early adopters were more likely than middle or late adopters to have freestall housing.

A similar methodology was used by Kinnucan et al. (1990) in 1984 to question dairy farm managers in Alabama, Florida, Georgia, and Mississippi. The survey document included a "Fact Sheet" but no fictitious advertisement, and was mailed to 1,000 randomly selected dairy farms; 313 (31%) returned usable responses. Comparison of respondent characteristics with the 1982 Census of Agriculture indicated the survey data tended to over-represent younger farm managers. Herd size of respondents was the only variable to differ significantly across the four states, averaging 95 head for Mississippi, 125 head for both Alabama and Georgia, and 533 head for Florida.

Awareness of BST among dairy farm managers was low, with 50 percent reporting not having heard of the technology before the survey (in 1984). Another 33 percent indicated little awareness of the product, 14 percent were somewhat familiar with it, and only 3 percent very familiar. The majority of respondents saw potential for this new technology, but 30 percent were either skeptical of its commercial feasibility or questioned the desirability of a product which could increase the supply of milk and lead to a reduction in milk prices. Forty-one percent indicated they would adopt BST immediately upon availability; another 36 percent said within one year. Only 8 percent said they would never use BST.

Of the potential adopters in the Kinnucan et al. (1990) study, 66 percent indicated they would begin using BST on only a few cows, 15 percent on half the herd, and 19 percent would initially use the product on the entire herd. Willingness to purchase BST was very sensitive to its cost, with 95 percent of all respondents willing to purchase at a daily cost of 10 cents per cow, but only 2 percent would purchase at a cost of 40 cents per cow per day.

Kinnucan et al. (1990) used a logit analysis to relate manager and farm characteristics to level of awareness of BST and the decision whether or not to adopt. Older, more-educated dairy farm managers with larger herd sizes tended to be more aware of BST before the survey. Factors positively correlated with rapid adoption were herd size and the use of artificial insemination, while output per cow and use of freestall housing were negatively correlated with rapid adoption. Output per cow and the relative importance of dairying as a source of income were negatively related to the extent of initial use in a herd. This suggests that farms relying more on milk for income, and herds with greater output per cow, were less willing to use BST on their entire herd without first experimenting on a few cows.

The only other pre-1986 research regarding potential adoption and diffusion of BST was reported by the Office of Technology Assessment (1985). Instead of surveying dairy farm managers, the Office of Technology Assessment (OTA) questioned leading agricultural scientists about the potential impact of emerging technologies, including BST, on agricultural productivity. A Delphi process was used to project the impacts of the various technologies for the years 1990 and 2000. OTA reported that these experts expected BST to be rapidly adopted in the dairy industry due to the clear economic advantages anticipated for the technology. A farm-level analysis in the same report, despite assertions by leading scientists that BST is a size-neutral technology (Kalter and Milligan 1987), OTA assumed that only those farms in the largest herd-size category in each of three dairy regions of the country would adopt BST immediately following commercial introduction.

Relying on these studies, BST adoption and diffusion were expected to be more rapid than that of past technologies made available to dairy farm managers. This assessment was based largely on scientific data from BST-research trials available at that time, and then questioning scientists or surveying farm managers provided with this data. In 1985, references to BST began appearing more frequently in scientific journals, farm publications, and even in the popular press.

Post-1985 Studies on Potential Adoption of BST

The editors of *Dairy Herd Management* (*DHM*) magazine in 1986 mailed a survey to 954 dairy farm managers randomly selected from its subscription list (Annexstad 1986). Of the 26 percent that responded, 218 (88%) had heard of BST before receiving the survey. Of those, 62 percent indicated at least some willingness to adopt BST and only 15 percent said they would never use it. While a statistical analysis was not presented, a higher percentage of farm managers with high output per cow (> 18,000 lb/year) planned to use BST than did farm managers with low output per cow (< 15,000 lb/year). Dairy farm managers with the largest herd sizes (>500 head) were also more willing to adopt than were those having the smallest herd sizes (<40 head). Age of farm manager was a factor; the very youngest and very oldest groups

(18 - 34 years and 55 years or older, respectively) expressed a greater reluctance to adopt BST than did middle-aged managers.

When asked how they would manage BST, 73 percent of those who plan to use the technology will produce more milk with the same number of cows. An additional group (21%) indicated they will produce the same volume of milk using BST with fewer cows. The remaining respondents (6%) will milk more cows and produce more milk per cow. One-half of the respondents indicated they would need a rate of return of three to one in order to adopt; the other half said two to one would be sufficient incentive.

Respondents in this *DHM* survey also expressed some concerns about the use of BST. Eighty-three percent were concerned, to some degree, about possible adverse effects on cows; only 9 percent were unconcerned about this. Most of the respondents also had an interest in the image of dairy products held by consumers; 82 percent were concerned about the possible impact of BST, and only 6 percent were unconcerned about this.

A 1987 personal-interview survey to assess Wisconsin dairy farm residents' levels of knowledge of BST was conducted for the *Wisconsin Agriculturalist* (Marion et al. 1988). Of the 400 farm interviews (200 men and 200 women), 270 received at least part of their gross farm income from dairying. Fifty-seven (21%) of these 270 dairy farm residents either could not or would not answer 17 questions developed by University of Wisconsin dairy scientists about BST technology. Correct responses for individual questions ranged from 7 to 94 percent; the average across all questions was 65.5 percent correctly answered.

This study used rating scales to aggregate respondent answers about predicted adoption into a single measure based on three different methodologies (behavioral intent, farm characteristics, and behavioral consistency). This procedure predicted that 9 percent of the respondents were likely to be early adopters of BST. Another 41 percent were described as having an intermediate likelihood of adoption; 23 percent were not likely to use BST. While no reasons were given, more than 27 percent of the respondents (74 out of 270) were not given a potential-to-adopt score. Therefore, of those respondents having a potential-to-adopt score, 69 percent were considered likely to adopt BST.

Those Wisconsin respondents classed as likely to adopt (either early or intermediate) had more years of formal education, were younger, gave more correct answers to the BST knowledge questions, and were more supportive of the University of Wisconsin's research on BST than were improbable adopters (all statistically significant). With respect to farm characteristics, likely adopters employed more nonfamily labor, had higher gross farm incomes, sold more total milk, and produced more milk per cow than did the improbable adopters (all statistically significant).

All respondents were asked which of three areas--capital, management or labor--would constrain their decision to adopt BST. Labor was the most likely constraining factor, especially if more nonfamily help would be required, while management was the least-constraining factor. When correlated with measures of farm scale, these constraints challenge the assertion that BST is a size-neutral technology made (Kalter and Milligan 1987). Larger herds tend to have more total milk sales, and thus were less likely to be constrained by capital, but more constrained by the need for nonfamily labor. In addition, larger herds with more nonfamily labor, and all herds with higher average output per cow anticipated fewer management constraints in adopting BST than other farms.

Zepeda (1989) reported a 1987 telephone survey of 153 California Grade A dairy farm managers. A response rate of 86 percent was achieved for those dairies still in business and willing to participate in the survey. Despite widespread press reports in that state of the controversy surrounding BST, 21 percent of respondents had not heard of BST prior to the survey. Of the 79 percent familiar with BST, only 8 percent indicated they would use the product as soon as it became available; 34 percent would wait before adopting, and 29 percent would not use BST (the remaining 8% were undecided).

Zepeda (1990) used this same data in a technology-adoption model to test hypotheses about BST adoption and forecast adoption patterns in California. Factors found to increase the probability of early adoption were more years of formal education, younger farm managers, more involvement in the dairy industry, and larger herd sizes. Those with a cautious, but receptive, attitude towards BST tended to be older managers and had fewer cows, but higher output per cow. The probability of a farm manager not using BST increased with the manager's age and years of formal education, but decreased with larger herd sizes.

This California study also questioned dairy farm managers about their concerns regarding BST (Zepeda 1989). Those who would not use BST were concerned most about consumer reaction (39%), while 29 percent were worried about the effects of BST on cow health.

Twenty-four percent would not use the product because their milk handlers had indicated they would not accept milk from cows supplemented with BST. Of the potential users of the technology, 38 percent expressed concern that BST would adversely affect milk-price levels, and 28 percent were concerned about consumer reaction. Twenty-three percent of those who would use BST thought that milk from BST-supplemented cows might not be safe.

A 1988 mail survey of dairy farms in several northern dairy states included a question about farm plans for adoption of BST (Borton et al. 1990). Responses from dairy farm managers in Indiana, Michigan, Minnesota, Missouri, Pennsylvania, and Wisconsin were summarized. The number of managers answering this question ranged from 144 in Missouri to 782 in Minnesota. On average across the six states, 9 percent of all respondents indicated they would adopt the product immediately; the lowest probable immediate-adoption response (6.9%) was in Missouri and the highest (9.8%) in Michigan. On the other hand, 46 percent of the respondents said they would not adopt BST, ranging from 40.7 percent in Indiana to 52.5 percent in Wisconsin.

The other 45 percent of all respondents in this survey indicated they would adopt BST, but only following successful use by neighbors, research reports and/or recommendations by university faculty, or reports of research by industry groups. While statistical analysis was not presented, Borton et al. (1990) reported that plans for adoption of BST as percent of milk volume was greater than planned adoption in terms of percent of farms. This indicated that farm managers with greater milk sales, from having larger herds and/or higher output per cow, were more willing to adopt the product than other farm managers.

Research on Potential BST Diffusion

Only two studies reported potential BST diffusion rates based on survey data. In the first study of potential BST adoption, Kalter et al. (1984) projected diffusion using three data subsets; all respondents, only those providing complete responses, and excluding those indicating only partial herd-adoption by year ten. Adoption always exceeded 40 percent of farms by the end of the first year, and ranged from 55 to 90 percent after 10 years. Kinnucan et al. (1990) found that the adoption curve tended to be S-shaped, with BST diffusion reaching 89 percent of all cows after 10 years.

Other researchers based projections of diffusion on the economics of using BST with current or forecasted costs and returns to producing milk. Fallert, et al. (1987) reported diffusion rates based on net returns to BST use under four milk price/policy scenarios. The percent of all farms adopting the technology ranged from 10 to 12 percent by the end of year one following commercial availability, and 45 to 70 percent by year 7, depending on scenario assumptions. An adoption rate for Wisconsin, described by Marion et al. (1988) as optimistic, projected use by 7 percent of the state's dairy farms at the end of the first year of availability, increasing to 40 percent by the end of year four.

A number of other BST-impact studies based potential diffusion on one or more of the studies cited above or other subjective criteria. At one extreme, Boehlje et al. (1987) assumed 100 percent adoption in the first year of commercial availability. Schmidt (1989), on the other hand, expects that "...a much more realistic estimate of the cows receiving BST would be 20 to 30 percent with the population of cows limited to higher producing cows." In a recent report to Congress, U.S. Department of Agriculture (1991) for the first time included the assumed commercial introduction of BST in a baseline forecast of U.S. milk supply and demand. This baseline projection assumed BST would be commercially available in 1992, and within 10 years would be used on 90 percent of all second lactation and older cows.

Summary

The rate of adoption is a key determinant in projecting the potential impacts of a new technology. Factors affecting the rate of adoption of any technology include its relative economic advantage, compatibility with existing resources, complexity and cost, divisibility, and communicability. Individual dairy farm managers will evaluate BST in terms of its characteristics and constraints with respect to each farm's objective function. The dairy industry has tended to embrace new technologies in the past, although few innovations achieved 100 percent adoption.

The emerging public policy debate on agricultural applications of biotechnologies in general and BST in particular demand *ex ante* impact assessment of such products. A key variable in such assessment is the speed and extent to which emerging products will be adopted by farms. Methods to obtain estimates of patterns of adoption of an innovation prior to its commercial availability rely on surveys of potential adopters' behavioral intent, constraints to adoption, and behavioral consistency in using previously developed technologies. A fourth method is to seek expert opinion on potential rates of adoption and diffusion.

Surveys of dairy farm managers about potential adoption of BST are summarized in Table 7.2. All but one study indicates that a majority of farm managers are likely to adopt the product within ten years after commercial availability. The earliest reported research on farm managers' intent to adopt was based on information prepared for and provided to the respondents. Such information was based on dairy science research to date, and the studies were done largely before appearance of significant reports in agricultural publications or the mass media. These studies predicted fairly rapid and widespread adoption of BST. Later (post-1985) research revealed a smaller percentage of likely adopters, and more concerns among dairy farm managers about BST impacts on cow health, consumer reaction, and potential constraints to adoption. Even more significant is the increase, over time, in the percent of survey respondents claiming they would not adopt BST.

Since it requires no explicit capital investment by farm managers, BST is often cited as a scale-neutral technology. However, most research on dairy farm managers' intent to adopt BST, especially early adoption, indicates that larger herd size increases the probability of adoption. Other farm and manager characteristics positively influencing the adoption decision include more years of education, younger age of manager, and use of previously introduced technologies such as DHI, AI, and freestall housing. The relationship between likelihood of BST adoption and average output per cow appears mixed.

It appears that BST will be viewed as an additional management tool that dairy farm managers may use to boost farm productivity and profitability. One potential advantage of this technology is its applicability to be used on only part of a herd initially, allowing dairy farmers to experiment with BST within their own production and management systems. Respondents in most surveys indicated that this will be their approach in deciding whether or not to adopt BST.

TABLE 7.2. Summary of BST Adoption and Diffusion Research.

Study	*Survey* *Year*	*Survey* *Area*	*Farm Managers Who Would Adopt*	*Farm Managers Who Would Not Adopt*
			- - - percent	- - -
Kalter et al. (1987)	1984	NY	61	13
Kinnucan et al. (1990)	1984	AL,FL,GA,MS	77	8
Annexstad (1986)	1986	*DHM* Readership	62	15
Marion et al. (1988)	1987	WI	50	23
Zepeda (1989)	1987	CA	42	29
Borton et al. (1990)	1988	IN,MI,MN,MO,PA,WI	54	46

References

Annexstad, J. 1986. "Survey says you'll use somatotropin, have concerns." *Dairy Herd Management* 23(10):13-15.

Bauman, D. E., M. J. DeGeeter, C. J. Peel, G. M. Lanza, R. C. Gorewit, and R. W. Hammond. 1982. "Effect of recombinantly-derived bovine growth hormone (bGH) on lactational performance of high yielding dairy cows." *Journal of Dairy Science* 65(Suppl 1):121(Abstract).

Borton, L. R., L. J. Connor, L. G. Hamm, B. F. Stanton, J. W. Hammond, M. Bennett, W. T. McSweeney, J. E. Kadlec, and R. M. Klemme. 1990. *Summary of the 1988 Northern U.S. Dairy Farm Survey*. East Lansing, MI: Michigan Agricultural Experiment Station. Research Report 509.

Boehlje, M., G. Cole, and B. English. 1987. "Economic impact of bovine somatotropin on the U.S. dairy industry." Pp 167-76 in *Proceedings from the National Invitational Workshop on Bovine Somatotropin*. St. Louis, MO. Sept. 21-23.

Buttel, F. H., and C. C. Geisler. 1987. "The social impacts of bovine somatotropin: emerging issues." Pp 196-208 in *Proceedings from the National Invitational Workshop on Bovine Somatotropin*. St. Louis, MO. Sept. 21-23.

Conneman, G. J. 1987. "Historical perspective on the adoption of new technologies on dairy farms." National DHIC Farmers' Biotechnology Conference. Philadelphia, PA: March 10, 1987.

Fallert, R. F., T. McGuckin, C. Betts, and G. Bruner. 1987. *bST and the Dairy Industry: A National, Regional, and Farm-Level Analysis*. Agr. Econ. Report No. 579. Washington, D.C: Commodity Economics Division, Economic Research Service, USDA.

Griliches, Z. 1957. "Hybrid corn: An explanation in the economics of technical change." *Econometrica* 25:501-22.

Just, R. E., A. Schmitz, and D. Zilberman. 1979. "Technological change in agriculture." *Science* 206:1277-80.

Kalter, R. J., R. Milligan, W. Lesser, W. Magrath, D. Bauman, A. McGuirk, E. Andrysick, and M. Grosh. 1984. *Biotechnology and the Dairy Industry: Production Costs and Commercial Potential of the Bovine Growth Hormone*. A.E. Research 84-22. Ithaca, NY: Department of Agricultural Economics, Cornell University.

Kalter, R. J., and R. A. Milligan. 1987. "Factors affecting dairy farm management and profitability." Pp 131-47 in *Proceedings from the National Invitational Workshop on Bovine Somatotropin*. St. Louis, MO. Sept. 21-23.

Kinnucan, H., U. Hatch, J. J. Molnar, and R. Pendergrass. 1990. *Adoption and Diffusion Potentials for Bovine Somatotropin in the Southeast Dairy Industry*. Auburn, AL: Alabama Agricultural Experiment Station Bulletin 605.

Koch, James V. 1980. *Industrial Organization and Prices*. Second Edition. Englewood Cliffs, NJ: Prentice-Hall, Inc.

Manchester, A.C. 1983. *The Public Role in the Dairy Economy*. Boulder, CO: Westview Press. p 49.

Marion, B. W., R. L. Wills, and L. J. Butler. 1988. *The Social and Economic Impact of Biotechnology on Wisconsin Agriculture*. Madison, WI: University of Wisconsin College of Agricultural and Life Sciences. Report prepared for the Wisconsin State Legislature.

Nowak, P. 1987. "The political and methodological context of social impact assessments: A case example of bovine somatotropin." Pp 209-33 in *Proceedings from the National Invitational Workshop on Bovine Somatotropin*. St. Louis, MO. Sept. 21-23.

Office of Technology Assessment. 1985. *Technology, Public Policy, and the Changing Structure of American Agriculture: A Special Report for the 1985 Farm Bill*. U.S. Congress. Report OTA-F-272. Washington, D.C: U.S. Government Printing Office.

Rogers, E. M. 1983. *Diffusion of Innovations*. Third Edition. New York, NY: Free Press.

Schmidt, G. H. 1989. "Economics of using bovine somatotropin in dairy cows and potential impact on the U.S. dairy industry." *Journal of Dairy Science* 72(3):737-45.

Swanson, L. E., S. M. Camboni, and T. L. Napier. 1986. "Barriers to adoption of soil conservation practices on farms." Pp 108-120 in *Conserving Soil: Insights From Socioeconomic Research*. Ankeny, IA: Soil Conservation Society of America.

U. S. Department of Agriculture. 1991. *Milk Inventory Management Programs*. Report submitted to The Committee on Agriculture, U.S. House of Representatives and to The Committee on Agriculture, Nutrition, and Forestry, U.S. Senate.

Zepeda, L. 1989. "Attitudes of California milk producers toward bovine somatotropin." *California Agriculture* March-April(1989):11-12.

Zepeda, L. 1990. "Predicting bovine somatotropin use by California dairy farmers." *Western Journal of Agricultural Economics* 15(1):55-62.

8

BST Impacts on Resource Needs and on Beef and Veal Output

George L. Greaser

Other chapters in this volume have documented that BST adoption by the nation's dairy farmers can be expected to have significant long-run consequences on milk output per cow, feed requirements per cow, total number of cows in the national herd, market prices of producer milk, and the size distribution of dairy farms. All this suggests, in turn, that BST adoption can be expected to have significant impacts on the dairy industry's production of nonmilk products and on its requirements for land and labor resources. The purpose of this chapter is to provide estimates of these latter impacts for each of eight regions in the United States. These estimates are based on farm-cost-and-return budgets developed for a typical dairy farm in each of the eight regions, and then aggregated to reflect total output and total input requirements for the respective regions.

Model Assumptions

The model uses many of the same assumptions and analytical approaches employed by Butler (Chapter 6). The basic methodology involves (1) developing a cost-and-return budget for an individual cow in the region of concern and for the specific BST scenario being examined, and (2) multiplying these results by the number of cows needed to produce the specified amount of milk in that region, under the assumed response to BST. The regional definitions, BST scenarios examined, output response to BST, price farmers can expect to receive for milk, and number of cows needed in each region given the assumed output per cow and regional demand requirements under each of the scenarios examined were taken from Fallert and Hallberg (Chapter 10). The results generated here are to be interpreted as the long-run consequences of BST adoption assuming the markets for milk and other products of the dairy industry, as well as the input markets, have fully adjusted to the changes in production brought about by BST adoption. The budgets generated for this analysis were developed as consistently as possible with the production and cost data reported by Economic Research Service (U. S. Department of Agriculture 1990).

Long-run equilibrium prices received by farmers for nonmilk products and prices paid by farmers for feed and other inputs were assumed to remain at the 1989 levels reported by the U. S. Department of Agriculture (1990), whether or not BST is adopted. This, of course, is inappropriate in that BST adoption can be expected to produce changes in veal and utility beef output and in feed and labor inputs, which in turn can be expected to lead to changes in equilibrium prices of these goods. Unfortunately, a full general equilibrium analysis is beyond the scope of the study reported here, so general equilibrium prices for the latter commodities were not available. Thus it was assumed that BST adoption would have no impact on the long-run price of veal, utility beef, labor, or feed. Based on related research in this area (see, e.g., Gum and Martin 1990) this appears to be a reasonable assumption, except possibly in the case of veal. In any event, this assumption will have little or no consequence on the results of principal concern to this chapter; the nonmilk output and input requirements of the dairy industry in an environment of general BST use.

It was assumed that BST will be administered via injections, that these injections will be administered at 14-day intervals beginning 65 days into the lactation and terminating 265 days into the lactation. The additional labor required to administer 14 BST injections and to capture the additional milk produced was assumed to be two hours per cow.

To determine the increased nutrient requirements for the additional milk produced, the National Research Council (1978) data on nutrient requirements was used. These data show that a dairy cow needs an additional 0.304 pounds of total digestible nutrients (TDN) and 0.082 pounds of crude protein (CP) to produce an additional pound of 3.5 percent fat-corrected milk. When the protein requirement for additional milk production is met the TDN requirement, in all probability, will also have been met. Thus all additional nutrients needed were assumed to be supplied by a ration that fulfills the protein requirement. To assure that the cow's intake capacity is not exceeded, a standard 16 percent concentrate mix was assumed to be used to provide the additional nutrients required. According to these assumptions, the nutrient density of the ration was increased with little change in the cow's total ration intake. These assumptions are consistent with those of previous researchers (Fallert et al. 1987). Finally, it was assumed that each cow treated with BST would respond with the same percentage increase in milk output regardless of its production level prior to BST treatment.

The cost of BST used in this analysis was 30 cents per day per cow. The actual cost of BST is, of course, as yet unknown, but estimates range as high as 75 cents per day per cow (see Butler, Chapter 6). The lower estimate was used here on the basis that in the longer run (the period of relevance for this study) market forces will drive the cost of BST down to the 30-cents-per-day estimate.

Model Results

A summary of the aggregate regional results from this analysis is presented in Tables 8.1 through 8.8. Aggregate results for the United States as a whole are given in Table 8.9. As in Fallert and Hallberg (Chapter 10), a base-run scenario is included, for comparison purposes, in each of the tables. The data for this base run scenario were generated on the assumption that BST is *not* adopted; milk prices, output per cow, feed and labor requirements, etc. listed in column 1 of each table are those projected to exist, at equilibrium, under the status quo. All other scenarios examined assumed BST was adopted as explained by Fallert and Hallberg. Costs and returns data are provided for completeness in these tables, although for the purposes of this chapter we are primarily interested in the information on bob calves produced (vealers, steer calves, etc.), utility beef produced, labor used, and land area required for production of the feed needed by the dairy herd. Feed costs shown here include corn, soybean meal, feed by-products, hay, corn silage, and pasture, plus ration preparation costs and $60 for BST purchase in those scenarios where BST was used. Total cash costs include, in addition to feed and other variable costs, a charge for general farm overhead, taxes, insurance, and interest expense. Net returns include returns from the sale of milk, bob calves, and cull cows over and above cash costs.

Profitability

The results of the analysis verify that adoption of BST by all of the nation's dairy farmers can be expected to lead to substantial changes in the dairy industry. A comparison of columns 1 and 2 in Tables 8.1 through 8.8 suggests that under full adoption of BST net returns per cow can be expected to fall by 28 percent in the Southern Plains, 74 percent in the Northwest, and an average of 39 percent in the remaining six regions. This is due in part to the increase in per-cow feed costs associated with the increased milk output as BST is used. More importantly, however, it is due to the fact that under full adoption of BST the price of milk (in the long run) is expected to fall (by 7 - 8% as shown in Tables 8.1 through 8.8) so that the gross value of the milk produced, though greater in volume, fails to increase sufficiently to offset the added feed and other cash costs.

One might wonder: Why would dairy farmers adopt BST if it means lower net returns per cow? The point is that, at the lower prices expected in the long run, farmers would be forced to adopt BST or suffer a greater reduction yet in net returns per cow. This is the "technology treadmill" at work as discussed in Fallert and Hallberg (Chapter 10).

Clearly these results, if realized, imply significantly reduced annual incomes for farm families on dairy farms, unless compensated with income from other sources. Hence added pressure will be placed on the small dairy farms if they are to survive in the longer term. Further, if the surviving dairy farmers in a given region are unable to expand production (i.e., increase herd size) sufficiently to make up for the production shortfall caused by farmers exiting the industry, that region will lose market share. This is certainly of interest from the perspective of aggregate farm income for the region. It could also have a negative effect on the performance of milk-marketing institutions in the region.

The results also show that if either the Upper Midwest or the Northeast regions were to ban BST use while all other regions encourage dairy farmers to use this technology, net returns per cow in all regions would fall from the base-run situation by a somewhat lower percentage than would be the case if all regions adopted BST. Nevertheless, it should be noted that producers in the region banning BST would experience a greater decline in net returns per cow than would producers in other regions. In this sense, producers in a region

TABLE 8.1. Summary Results for Northeast Region.[a]

	Base Run	BST Adopted Everywhere	BST Banned in U.Midwest	BST Banned in Northeast	BST Adopted Everywhere and Reduced Item Consumption
Milk price ($/cwt)	12.44	11.42	11.69	11.73	10.92
Production (lb/cow)	14,364	15,801	15,801	14,364	15,801
Feed costs ($/cow)	790	847	847	790	847
Total variable costs ($/cow)	1,280	1,419	1,419	1,280	1,419
Total cash costs ($/cow)	1,646	1,785	1,785	1,646	1,785
Net returns ($/cow)	366	244	287	264	165
Number of cows (1,000 head)	1,776	1,692	1,716	1,707	1,648
Bob calves sold (1,000 head)	977	931	944	939	907
Utility cows sold (1,000 head)	550	525	532	529	511
Labor required (1,000 hours)	56,824	57,538	58,329	54,616	56,042
Land required (1,000 acres)					
Corn	1,708	1,793	1,817	1,642	1,746
Soybeans	1,599	1,678	1,701	1,537	1,635
Hay	1,554	1,481	1,502	1,494	1,443
Silage	990	944	957	952	919

[a] ME, NH, VT, CN, MA, RI, NY, and PA.

TABLE 8.2. Summary Results for Southeast Region.[a]

	Base Run	BST Adopted Everywhere	BST Banned in U.Midwest	BST Banned in Northeast	BST Adopted Everywhere and Reduced Item Consumption
Milk price ($/cwt)	13.61	12.66	12.82	12.77	12.16
Production (lb/cow)	13,056	14,362	14,362	14,362	14,362
Feed costs ($/cow)	1,055	1,105	1,105	1,105	1,105
Total variable costs ($/cow)	1,547	1,678	1,678	1,678	1,678
Total cash costs ($/cow)	1,786	1,917	1,917	1,917	1,917
Net returns ($/cow)	227	137	160	153	65
Number of cows (1,000 head)	1,136	1,051	1,066	1,062	1,005
Bob calves sold (1,000 head)	625	578	586	584	553
Utility cows sold (1,000 head)	352	326	331	329	312
Labor required (1,000 hours)	61,321	58,836	59,710	59,456	56,275
Land required (1,000 acres)					
Corn	1,739	1,701	1,727	1,719	1,627
Soybeans	1,628	1,593	1,617	1,610	1,524
Hay	530	491	498	496	469
Silage	263	244	247	246	233

[a] DE, MD, VA, WV, NC, SC, GA, FL, AL, MS, LA, AK, and TN.

TABLE 8.3. Summary Results for Upper Midwest Region.[a]

	Base Run	*BST Adopted Everywhere*	*BST Banned in U.Midwest*	*BST Banned in Northeast*	*BST Adopted Everywhere and Reduced Item Consumption*
Milk price ($/cwt)	11.66	10.67	10.96	10.87	10.19
Production (lb/cow)	13,792	15,171	13,792	15,171	15,171
Feed costs ($/cow)	810	859	810	859	859
Total variable costs ($/cow)	1,168	1,295	1,168	1,295	1,295
Total cash costs ($/cow)	1,562	1,690	1,562	1,690	1,690
Net returns ($/cow)	281	164	185	195	92
Number of cows (1,000 head)	2,487	2,329	2,361	2,362	2,249
Bob calves sold (1,000 head)	1,368	1,281	1,299	1,299	1,237
Utility cows sold (1,000 head)	771	722	732	732	697
Labor required (1,000 hours)	47,246	48,905	44,866	49,596	47,229
Land required (1,000 acres)					
Corn	2,239	2,315	2,126	2,348	2,236
Soybeans	2,096	2,167	1,991	2,198	2,093
Hay	2,645	2,477	2,512	2,512	2,392
Silage	1,204	1,127	1,143	1,143	1,089

[a] WI and MN.

TABLE 8.4. Summary Results for Corn Belt Region.

	Base Run	*BST Adopted Everywhere*	*BST Banned in U.Midwest*	*BST Banned in Northeast*	*BST Adopted Everywhere and Reduced Item Consumption*
Milk price ($/cwt)	12.18	11.21	11.48	11.42	10.71
Production (lb/cow)	13,534	14,887	14,887	14,887	14,887
Feed costs ($/cow)	788	837	837	837	837
Total variable costs ($/cow)	1,197	1,325	1,325	1,325	1,325
Total cash costs ($/cow)	1,590	1,718	1,718	1,718	1,718
Net returns ($/cow)	294	187	227	218	113
Number of cows (1,000 head)	1,833	1,707	1,741	1,733	1,643
Bob calves sold (1,000 head)	1,008	939	958	953	903
Utility cows sold (1,000 head)	568	529	540	537	509
Labor required (1,000 hours)	45,823	46,084	47,007	46,789	44,348
Land required (1,000 acres)					
Corn	2,049	2,065	2,106	2,096	1,987
Soybeans	1,919	1,933	1,972	1,963	1,860
Hay	1,744	1,624	1,657	1,649	1,563
Silage	684	637	650	647	613

[a] MO, IN, IL, KY, IA, OH, and MI.

TABLE 8.5. Summary Results for Upper Plains Region.[a]

	Base Run	*BST Adopted Everywhere*	*BST Banned in U.Midwest*	*BST Banned in Northeast*	*BST Adopted Everywhere and Reduced Item Consumption*
Milk price ($/cwt)	12.29	11.33	11.36	11.38	10.81
Production (lb/cow)	13,611	14,972	14,972	14,972	14,972
Feed costs ($/cow)	788	837	837	837	837
Total variable costs ($/cow)	1,197	1,325	1,325	1,325	1,325
Total cash costs ($/cow)	1,590	1,718	1,718	1,718	1,718
Net returns ($/cow)	318	214	219	222	136
Number of cows (1,000 head)	644	607	606	607	588
Bob calves sold (1,000 head)	354	334	333	334	323
Utility cows sold (1,000 head)	200	188	188	188	182
Labor required (1,000 hours)	16,099	16,376	16,356	16,400	15,870
Land required (1,000 acres)					
Corn	720	734	733	735	711
Soybeans	674	687	686	688	666
Hay	613	577	577	578	559
Silage	240	226	226	227	219

[a] ND, SD, WY, MT, UT, CO, KS, and NE.

TABLE 8.6. Summary Results for Southern Plains Region.[a]

	Base Run	*BST Adopted Everywhere*	*BST Banned in U.Midwest*	*BST Banned in Northeast*	*BST Adopted Everywhere and Reduced Item Consumption*
Milk price ($/cwt)	13.17	12.22	12.44	12.38	11.73
Production/cow (lbs)	13,821	15,203	15,203	15,203	15,203
Feed costs ($/cow)	1,015	1,071	1,071	1,071	1,071
Total variable costs ($/cow)	1,431	1,568	1,568	1,568	1,568
Total cash costs ($/cow)	1,696	1,833	1,833	1,833	1,833
Net returns ($/cow)	360	260	294	285	186
Number of cows (1,000 head)	561	513	524	521	488
Bob calves sold (1,000 head)	308	282	288	287	268
Utility cows sold (1,000 head)	174	159	162	162	151
Labor required (1,000 hours)	20,187	19,486	19,918	19,798	18,531
Land required (1,000 acres)					
Corn	671	662	677	673	629
Soybeans	628	620	633	630	589
Hay	752	688	703	699	654
Silage	38	35	35	35	33

[a] NM, OK, and TX.

TABLE 8.7. Summary Results for Northwest Region.[a]

	Base Run	*BST Adopted Everywhere*	*BST Banned in U.Midwest*	*BST Banned in Northeast*	*BST Adopted Everywhere and Reduced Item Consumption*
Milk price ($/cwt)	10.99	10.17	10.31	10.23	9.68
Production (lb/cow)	18,096	19,905	19,905	19,905	19,905
Feed costs ($/cow)	1,093	1,178	1,178	1,178	1,178
Total variable costs ($/cow)	1,737	1,902	1,902	1,902	1,902
Total cash costs ($/cow)	2,037	2,203	2,203	2,203	2,203
Net returns ($/cow)	175	45	73	57	-52
Number of cows (1,000 head)	430	395	402	398	373
Bob calves sold (1,000 head)	236	217	221	219	205
Utility cows sold (1,000 head)	133	123	125	123	116
Labor required (1,000 hours)	23,196	22,144	21,708	22,298	20,886
Land required (1,000 acres)					
Corn	382	400	407	403	377
Soybeans	357	374	381	377	353
Hay	407	375	381	377	353
Silage	90	83	84	83	78

[a] OR, WA, and ID.

TABLE 8.8. Summary Results for West Coast Region.[a]

	Base Run	*BST Adopted Everywhere*	*BST Banned in U.Midwest*	*BST Banned in Northeast*	*BST Adopted Everywhere and Reduced Item Consumption*
Milk price ($/cwt)	11.84	10.90	11.11	11.05	10.47
Production/cow (lbs)	17,530	19,283	19,283	19,283	19,283
Feed costs ($/cow)	1,215	1,297	1,297	1,297	1,297
Total variable costs ($/cow)	1,666	1,824	1,824	1,824	1,824
Total cash costs ($/cow)	2,032	2,190	2,190	2,190	2,190
Net returns ($/cow)	268	136	177	165	53
Number of cows (1,000 head)	1,343	1,206	1,235	1,226	1,147
Bob calves sold (1,000 head)	739	663	679	674	631
Utility cows sold (1,000 head)	416	374	383	380	356
Labor required (1,000 hours)	61,776	57,887	56,804	58,831	55,079
Land required (1,000 acres)					
Corn	1,048	1,084	1,110	1,102	1,032
Soybeans	981	1,015	1,039	1,032	965
Hay	1,060	952	975	968	906
Silage	280	252	258	256	240

[a] CA, AR, and NV.

TABLE 8.9. Summary Results for the United States.

	Base Run	*BST Adopted Everywhere*	*BST Banned in U.Midwest*	*BST Banned in Northeast*	*BST Adopted Everywhere and Reduced Item Consumption*
Number of cows (1,000 head)	10,210	9,500	9,651	9,616	9,141
Bob calves sold (1,000 head)	5,616	5,225	5,308	5,289	5,027
Utility cows sold (1,000 head)	3,164	2,946	2,993	2,980	2,834
Labor required (1,000 hours)	332,472	327,256	324,698	327,784	314,260
Land required (1,000 acres)					
Corn	10,556	10,754	10,783	10,718	10,345
Soybeans	9,882	10,067	10,020	10,035	9,685
Hay	9,305	8,665	8,805	8,773	8,339
Silage	2,585	2,421	2,457	2,446	2,335

banning BST would be disadvantaged. They would be further disadvantaged, to the extent that their region would lose market share.

Bob Calves and Utility Beef Sold

The number of bob calves produced determines the number of veal calves and the number of finished dairy steers and heifers for slaughter. The dairy herd replacement rate determines the number of utility beef animals to be sold for slaughter. The number of both of these classes of animals is in turn affected by the number of cows necessary to supply the amount of milk needed under the various scenarios examined. The number of bob calves available for market was assumed equal to 55 percent of the number of cows in the herd. The number of utility animals available for slaughter was assumed equal to 31 percent of the number of cows in the herd. Thus the decline in the number of bob calves and cull animals sold is directly related to the decline in number of cows in the herd under each of the scenarios examined.

The West Coast and Northwest regions show the greatest declines (8-10%) in number of animals sold. Milk output per cow is the highest in these two regions, and a larger decline in number of cows will occur here as BST is adopted. More modest declines (4-5%) are projected in the remaining regions. The most dramatic changes are projected to occur if a reduction in demand for milk and dairy products accompanies adoption of BST in all regions.

Labor Use

The effects of BST adoption on labor required varies greatly from region to region. Since labor use is such a small part of producing milk, however, the changes in this input are not great. Under some of the scenarios examined, the additional hours of labor required when BST is adopted increases the total labor requirement slightly in the Northeast, Upper Midwest, Upper Plains, and Corn Belt regions. In other regions (the Northwest, West Coast, Southern Plains, and Southeast, where labor costs are a somewhat more significant part of milk production), the projected change in number of cows is sufficient to bring about a reduction in total labor use from 4 to 6 percent when BST is adopted in all regions. When a reduction in demand for milk and dairy products accompanies BST adoption, labor use is reduced by as much as 10 percent in some regions.

Hay and Haylage

The amount of hay and haylage required declines in all regions and under all scenarios. The assumption that nutrient density of the ration be such that all nutrient requirements of the cow are met without exceeding its feeding capacity requires that the percentage decline in the amount of hay and haylage needed is exactly equal to the percentage decline in the number of cows. To determine the acreage needed to provide the hay requirement in each region and for each scenario examined, the hay requirement was divided by the 1989 average yield in the respective region.

If BST is adopted everywhere, in some regions as much as 10 percent of the present hay acreage would no longer be needed by the dairy enterprise and would be idled or diverted to other enterprises such as beef, sheep, or field crops. Clearly, not all of this acreage can be diverted to row crops. Some is not suited for anything but hay or pasture. Further, some hay acres may be retained for rotation cropping and/or conservation purposes.

Corn and Soybean Acres

The amount of corn silage, corn grain, and soybeans needed are also tied proportionally to the number of cows needed. Thus, as the number of cows change under the various scenarios examined, so too does the acreage needed for corn silage, corn grain, and soybeans. To determine the acreage needed for corn silage in each region and for each scenario, the total silage requirement was divided by the 1989 average per-acre corn silage yields for the respective region. The corn and soybean acreages needed to meet the nutrient requirements for the different regions and scenarios was determined by dividing the total grain requirement by the most recent three-year weighted average corn and soybean per-acre yield in the respective region. If the estimated corn or soybean requirement exceeded that available from the local crop acreage (as determined by the maximum crop acreages of each crop over the most recent three-year period), a weighted-average yield for the nation was used to estimate the additional acreage needed.

The projections show that, when BST was adopted everywhere, corn and soybean acreages increased for every region except the Southeast and Southern Plains. In the Southeast and Southern Plains, the percent of grain in the ration was much higher than in the rest of the country, thus any change in the number of cows affected the amount of grain needed more here than in other regions. Corn and soybean acreage requirements declined in those regions banning BST because these regions lost some of their market share. Banning BST brought about a reduction in number of cows, and since output per cow did not increase in these regions, corn and soybean requirements also decreased. When a demand reduction was projected with adoption of BST in all regions, grain acreages required decreased except in the Northeast. Apparently, because the bulk of the nation's population lives in or near the Northeast region, this region will continue to produce the product (fluid milk) that is more expensive to ship, and thus this region losses less of its market share than do other regions.

Summary

The long-run consequences of BST adoption are likely to be more dramatic on net revenue to dairy farmers than on the nonmilk products of dairying or on the resources required by dairy farms. Nevertheless, some important changes in the latter will occur as the number of cows needed to meet the nation's demand for milk falls, as projected with BST adoption.

For the nation as a whole and under BST adoption in all eight regions, the projected 7 percent decline in number of cows means that veal calves and utility dairy cows sent to market will decline by a like percentage. The dairy industry now supplies an estimated 25 percent of the nation's utility beef. This would mean a 1.5 percent decrease in dairy beef available for the utility beef market. Such a decrease is not likely to have much impact on beef prices. On the other hand, the dairy industry supplies about 80 percent of the veal market in the United States. Thus a 7 percent reduction in dairy calves can be expected to lead to a reduction of more than 5 percent of the veal supply. This would have a much more noticeable impact on the veal market, and might add a modest amount (say, $3 - $5) to the net returns per-cow estimates shown in columns 2 through 5 of Tables 8.1 through 8.8.

One might also expect a 7 percent reduction in number of dairy cows and calves to lead to a reduction in the incidence of pollution from animal manure and to a reduction in the amount of commercial fertilizer needed. If, however, we consider the increased feed required by the reduced number of dairy cows, the impact here is similarly not likely to be great. This conclusion is consistent with that of other researchers (Preckel et al. 1990) although, clearly, in specific areas with large declines in cow numbers the impact could be quite substantial.

Full adoption of BST in all regions is projected to lead to increases of slightly less than 2 percent in corn and soybean acres required for the United States as a whole. Again this would have a minimal impact on the prices of corn and soybeans. Full adoption of BST in all regions is also projected to lead to reductions in needed hay and silage acreages of 6.8 and 6.4 percent, respectively. Much of this acreage would likely be idled, unless other animal enterprises can be substituted for dairy. The corn silage acres not required for milk production cannot all be used for corn grain production because in some regions the growing season is not long enough for corn to ripen.

Full adoption of BST in all regions is projected to lead to a reduction in total labor hours required for milk production in the nation as a whole of 2 percent--the equivalent of about 1.5 or 2 man years. This would have practically no impact on the national labor market. Even under the worst-case scenario when reduced demand accompanies BST adoption in all regions, total labor required by the nation's dairy farms would decline by only about 5 percent. Nevertheless, in some regions and under certain scenarios, the labor impact could be substantial--especially in the Southeast and Northwest if reduced demand accompanies BST adoption in all regions.

Acknowledgments

I am grateful to Mr. Alton Royer for his assistance with the research upon which this chapter is based, and to Dr. Earl J. Partenheimer, Pennsylvania State University, and Dr. L. J.(Bees) Butler, University of California at Davis, for their helpful review comments.

References

Fallert, R. F., T. McGuckin, C. Betts, and G. Bruner. 1987. *bST and the Dairy Industry: A National, Regional, and Farm-level Analysis*. AER-579. Washington, D.C: Economic Research Service. USDA.

Gum, R. L., and W. E. Martin. 1990. *Economic impacts of biotechnical innovations in the U.S. and Arizona dairy and cotton industries*. Bulletin No. 267. Tucson, AR: Arizona State University Agricultural Experiment Station.

National Research Council. 1978. *Nutrient requirements of dairy cattle*. 5th Edition. Washington D.C: National Academy of Sciences.

Preckel, P. V., R. F. Turco, M. A. Martin, and C. H. Noller. 1990. "Introduction of bovine somatotropin: Environmental effects." Staff Paper 90-13. Lafayette, IN: Purdue University Department of Agricultural Economics. Staff Paper 90-13.

U. S. Department of Agriculture. 1990. *Economic Indicators of the Farm Sector: Costs of Production--Livestock and Dairy, 1989*. ECIFS 9-1. Washington D.C: Economic Research Service. USDA.

9

Impact of BST on Small Versus Large Dairy Farms

Loren W. Tauer

In the public debate on bovine somatotropin, three major issues typically surface: does BST alter milk quality, does it adversely affect herd health, and will it lead to the demise of the small dairy farm. The former two of the three issues are discussed elsewhere (Chapters 5 and 12). The purpose of this chapter is to discuss the impact of BST on small versus large dairy farms.

Some of the discussion of BST and small versus large dairy farms hinges on the ethical discussion of small versus large farms in U.S. society. Since there are often tradeoffs between small versus large farms, disagreement is often due to different values individuals place on outcomes. These issues are discussed by Thompson (Chapter 2). Yet, even if agreements are reached on values, disagreements can still occur concerning the impacts of BST use. This conflict may occur because of illogical reasoning of the impact of BST on farm size. Discourse may also include disagreements concerning per-cow yield increases and adoption rates by different sizes of farms. My purpose is to develop a conceptual framework in which we can logically discuss the impact of BST on various farm sizes and then the impact of BST on farm-size distribution.

The relationship of BST to farm size hinges on whether small farms will adopt BST to the same extent as large farms and be able to use it as effectively. To a large extent, the adoption of BST will depend on how well a farmer can benefit from its use, although noneconomic factors may also influence the decision to adopt. Therefore, the starting point is to determine if there is something inherent in the use of BST that favors the large over the small farm. The next step is to determine if more larger than smaller farms will adopt BST, regardless of whether or not the larger farm gains more from the use of BST.

After the individual farm benefits are determined and the rate of adoption is resolved, it is necessary to discuss the market effects. Recombinant BST is an output-increasing technology. As such, it will decrease the price of milk, unless the government intervenes permanently by purchasing and disposing of the excess milk production. If milk price is not allowed to fall and farmers increase production, inputs used in dairy production may increase in price. Small farms may be more adversely affected by a milk-price decrease or input-price increase compared to large farms, even if they benefit from BST to the same degree and adopt BST to the same extent.

Benefits of BST in Relation to Farm Size

The effect of BST is to increase milk production from the cow, requiring additional feed and other inputs to produce that additional milk. Is there anything inherently different about large farms that would mean they could produce more milk per cow from BST, possibly using fewer inputs? Previous chapters in this volume have discussed the input and management requirements necessary to successfully use BST. Many scientists argue that the use of BST should not require the use of any equipment or practice that is not readily available to the smaller as well as the larger farm. The smaller farm can be just as effective in its herd health, breeding, and feeding programs, and introducing BST into the herd will not put the smaller farm at a disadvantage in herd management. Thus, there should be no per-cow yield difference as long as both large and small farms are using cows with identical milk-production potentials since, low milk producing cows may not respond as favorably to BST treatment as high milk producing cows. This was demonstrated by Marsh, Galligan, and Chalupa (1988) who simulated operating results for three milk-production levels and three herd sizes and showed that production per cow is more important than herd size in determining BST-use profitability per cow. In fact, medium- to high-producing herds benefit almost similarly, regardless of size. Even a medium-production farm with 32 cows could compete favorably.

The situation is different, however, if the productive capacity of cows on smaller farms is less than that of larger farms. Advocates of BST use state that this is not necessarily so. They state that there are many smaller herds with cows that are more productive than cows from some larger herds. That is quite likely true. But it is also true that, on average, smaller farms have lower-producing cows than larger farms.

A 1989 Wisconsin Agricultural Statistics Service survey reported that average milk production increases from 11,000 pounds per year for herds less than 10 cows, to 13,899 pounds per year for herds with 31 to 40 cows, to 15,934 pounds per year for herds with 51 to 60 cows (Marion and Wills 1990, Table 5). For herds of more than 150 cows, the average is 17,128 pounds per year. More significant is that in only 30.6 percent of the herds of 31 to 40 cows does average production exceed 16,000 pounds per year, while in 50 percent of the herds of 51 to 60 cows, average production exceeds 16,000 pounds per year. It is not known if the lower yields on the smaller farms are due to management limitations on those farms or to the limited genetic potential of the animals. It may be that both factors are responsible. Regardless, successful use of BST requires productive cows and farmers with good management skills.

These data imply that although many smaller farms may use BST as effectively as most larger farms, on average, smaller farms will not. Since BST use will increase aggregate milk production and thus require fewer cows in total, more of those reductions may come from smaller farms that cannot as effectively use BST than from larger farms. Since there will be fewer cows, many small farms may be eliminated before milk demand and supply are balanced.

There are those who argue that since BST increases production per cow, farmers will reduce the number of cows in their herd and produce the same amount of milk. If all farmers reduce their herd size accordingly, there would be no need for a reduction in farm numbers. This line of reasoning is economically illogical, and empirical evidence does not support it. It rarely makes economic sense for farmers to reduce the number of cows they have capacity for on their farms. Historically, when farmers have increased milk production per cow, they have not simultaneously decreased the average size of their herds. Why would they do that with BST use?

Adoption of BST by Farm Size

During the 1950s, a lively debate occurred between Griliches (an economist) and Brander and Straus, and Havens and Rogers (sociologists) concerning the determinants of technology adoption by farmers, especially for hybrid corn use. Griliches (1960) argued that the decision to adopt was an economic decision; Brander and Straus (1959), and Havens and Rogers (1961) argued that the decision was more a factor of characteristics of the farmer, location, and technology. Griliches (1962) countered that these factors indeed determine adoption because they determine the economics of the decision.

Current approaches state that the eventual decision to adopt or not indeed depends upon the economic benefits, but the timing of adoption depends upon characteristics of the farmer and the technology, since these determine the length of the learning process before the decision is reached. Within this context, a number of surveys asked farmers if and when they will adopt BST. Since BST is, of course, not yet available for commercial use, the questions were strictly hypothetical.

The first survey of this type, completed in 1984, found that those New York farmers indicating they would adopt BST early (within the first year) had herds averaging 72 cows in number. In contrast, herds of those farmers who said they would wait five years or longer or would never try BST averaged 49 cows (Lesser, Magrath, and Kalter 1986).

A 1986 survey of Georgia farmers found that those farmers who would adopt BST within one year had herds averaging 123 cows in size, while herds of those farmers who may adopt BST after five years or who may never adopt it averaged 110 cows (Carley, Fletcher, and Alexander 1989). A 1987 survey of California milk producers found that herd size had little effect on whether or not the producer was informed about BST (Zepeda 1990). However, 24 percent of producers with herds larger than 1,000 cows said they would use BST immediately upon availability, while only 1 percent of the queried producers with herds of fewer than 100 cows would use it immediately. In fact, 43 percent of those with 100 cows said they would never use it; only 26 percent of those with 1,000 cows indicated such a flat rejection.

These three surveys, in three different geographical regions of the United States, give consistent results. The managers of larger dairy farms, as a group, have indicated that they would adopt BST and adopt it sooner than would those with smaller dairy farms. Yonkers (Chapter 7) discusses and analyzes these and other surveys. Although their numerical magnitudes differ, these surveys indicate that the adoption of BST will be swift, with larger farms generally adopting BST before the smaller farms. Many scholars question the large and fast adoption rates suggested by these surveys, because such rates are unprecedented in comparison with adoption of previous dairy technologies (Schmidt 1989).

A potential limitation to these *ex ante* surveys is that what a farmer says he will do may be quite different from what he will do or what he must do in order to survive. Any increased aggregate milk produced from BST would put downward pressure on milk prices. This may force some who indicated they will not adopt BST to actually do so. Others, because of managerial limitations, will not use BST (although some may try it) because they cannot benefit from it. If we have a BST-differentiated market, some may also be able to receive a higher price for their non-BST-produced milk. However, even if non-BST milk does not command a higher price, the use of BST is not necessary for survival. Magrath and Tauer (1988), using a linear programming sector model, found that producers with medium- to low-producing herds can profitably survive without the use of BST. In fact, it will not be profitable for some of them to use BST. The cost of BST, additional feed, and other required inputs will exceed the value of the additional milk produced.

The Synergism

If smaller farms do not adopt BST to the same extent as larger farms (possibly because they do not benefit to the same extent from BST), then a logical conclusion is that BST will more negatively affect smaller farms than larger farms, because BST would increase aggregate milk output and lead to a milk-price decrease. The degree to which the milk price will fall depends upon numerous factors, discussed by Fallert and Hallberg (Chapter 10). If milk prices are not allowed to adjust downward, then the profitability of BST use will be bid into assets specific to dairy production. The extreme case of this is milk-production quotas. Depending upon market structure, the result on profitability may be similar to a milk-price reduction except that wealth effects occur from asset-price increases.

If a farmer does not adopt BST and the price of milk falls, his profits will fall. What is not clear to many people is that the profits of the BST adopters will likely also be less than would be the case if no farmer had the opportunity to use BST. Figure 9.1 shows such an occurrence from research completed by Tauer and Kaiser (1991). The figure shows how early adopters of BST earn greater profits, per cow, during the first four years than if BST were not available. However, after those first four years, their profits are lower than their profits would have been without the availability of BST. The question that is often asked, then, is why should they continue to use BST? The figure also provides that answer. If an individual farmer discontinues using BST while others continue to do so, the farmer who discontinues would have even lower profits, as shown by the profits of nonadopters of BST in the figure. This phenomenon was first discussed by Cochrane (1958) in the late 1950s as the "technology treadmill." Since technological change increases output and lowers prices, a farmer must adopt new technologies in order to survive. Since continuous technological changes occur, a farmer must stay on the "technology treadmill" or fail to remain competitive. Not every farmer, however, is able to remain on the treadmill.

This figure shows that nonadopters earn less profits. Earning lower profits has an impact on survivability. Even more importantly, however,

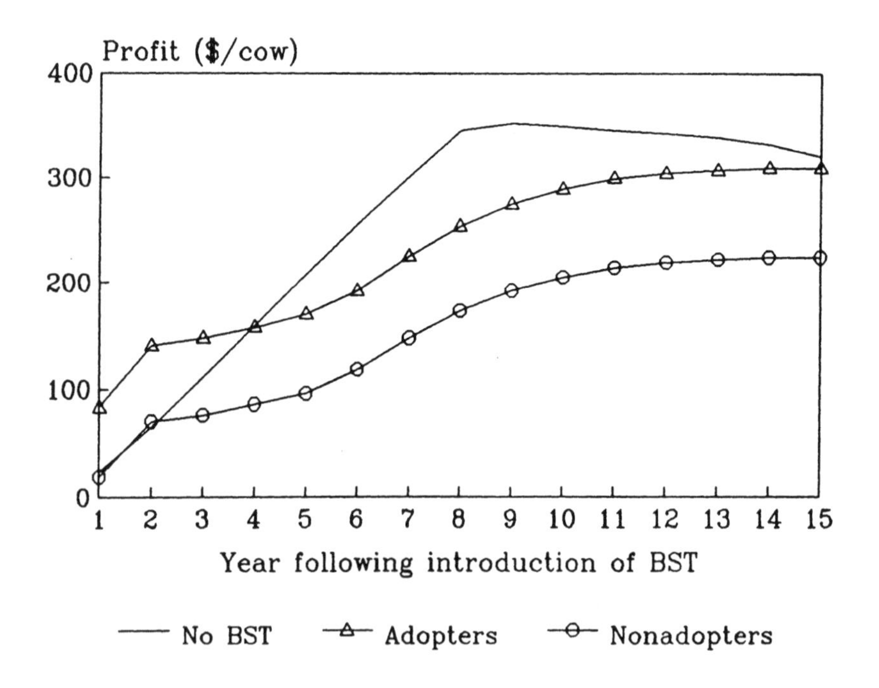

Figure 9.1. Anticipated profits ($/cow) for users and nonusers of BST, for fifteen years following introduction of BST, compared with profits anticipated if BST is unavailable to any producer.

the figure shows that significant adoption of BST will reduce the profits of all farmers, whether or not they adopt BST. Even if smaller farmsadopt and benefit from BST to the same extent as the larger farms, the same reduction in net profits per cow may place them in financial jeopardy simply because they have fewer cows.

Stanton and Bertelsen (1989) recently looked at operating results for U.S. dairy farms by size, using data from the 1987 National Farm Costs and Returns Survey. When comparing rates of return to farm assets, they found that once milk sales reach the $60,000 - $89,999 level (36-53 cows), the percentages are surprisingly similar, except for the largest size group of over $1,000,000 (600 plus cows). The rate of return is about 5 percent on farms with sales of less than $60,000, 10 percent for sales of $60,000 to $1,000,000, and 16.4 percent for sales in excess of $1,000,000 (Stanton and Bertelsen 1989, Table 16). Farms with sales exceeding $1 million currently produce 15.8 percent of all milk in the United States. Although the market share of that group will undoubtedly increase over time, the use of BST by that group will not, by itself, put many of the medium-size dairy farms out of business. At greatest risk are those farms with sales of less than $60,000 (fewer than 36 cows). Although collectively producing only 11.9 percent of the milk, this group includes 41.1 percent of the dairy farms. There are undoubtedly farms in the group that are competitive with any farm, regardless of its size, but as a group these small farms are most vulnerable. Any event that reduces dairy profitability will put pressure on these farms. The introduction of BST is such an event.

These smaller dairy farms are concentrated in the northern dairy states; the larger farms are more common in the West and Southwest. Case studies have consistently shown that the profitability of the larger herds in the Southwest is much greater than that of herds in the north (Office of Technology Assessment 1986). The implication is that in the future there will be more and larger dairy producers in the Southwest and fewer dairy farms in the northern states.

The limitation to these studies is that they do not incorporate the dynamics of the adjustment process. Dairy herd expansion or contraction in a region affects the demand for dairy-specific assets, which alters the price and value of these assets. No one believes that 100 percent of the nation's milk production will shift to California. So how much will shift? A recent article by Weersink and Tauer (1990) indicates that very little may shift. These economists used a dynamic regional production and demand model to forecast market shares into the future under different rates of technological change. They found that the West may continue to moderately increase production and market share, but that the traditional northern dairy states would lose very little market share. Nontraditional milk areas will produce less milk. These results are consistent with earlier results reported by Hallberg and Parsons (1986). The implication is that milk production may become even more concentrated into a few production regions, as has happened to other agricultural commodities.

Although the role of BST was not included in their analysis, Chavas and Magand (1988) looked at the size distribution of dairy farms in various states and regions of the United States. They found that economies of size, sunk costs, and market prices play a role in the evolution of the size distribution of dairy farms. Dairy farms with fewer than 50 cows, as a group, were found not to be efficient, and their numbers were projected to decline in all regions. Sunk costs of capital were found to provide some deterrent to rapid increases in farm size. These results imply that larger farms are more efficient, but past asset purchases by farmers prevent them from rapidly moving to larger sizes. Of most interest here, however, is the prediction that lower profitability would speed up the movement to larger farms. If BST adoption reduces profitability because of a lower milk price, then BST adoption will result in fewer small farms and more larger farms.

Bovine somatotropin is one of many different technologies that many believe increases the size and reduces the number of farms producing milk. In this chapter I have tried to reason through the process in which BST as a new technology may give larger farms a competitive advantage, such that they might increase in size or at least in number at the expense of smaller farms. BST is yet to be approved for commercial use, so its actual impact can only be conjectured. Other technologies, however, have been introduced into the dairy sector over the years, and it may be helpful to observe how they, as a group, have affected the numbers and average size of dairy farms.

Weersink and Tauer (1991) used state data from 1964 through 1987 to determine if there was a causality between increases in technology as measured by cow productivity and dairy farm size. Their results support only partially the view that technological change has caused increases in dairy farm size. Rather, the direction of causality appears to be due more to size than to the use of technology. This implies that new technology has not been a large factor in influencing larger farms, but that larger farms have led to the use of new technology. A specific example is that milking parlors have not led to larger farms, rather farms that wanted to become larger have used milking parlors to accomplish that. What appears to have been the driving force behind larger farms and the use of technology is prices. This supports the conclusions by Kislev and Peterson (1982) that reductions in farm numbers and increases in average farm size are due more to changes in relative prices than to technological change. The major determinant of changes in dairy farm structure may be dairy price policy.

Weersink and Tauer's (1991) empirical results on causality between productivity and average farm size would appear to contradict the analytical reasoning presented earlier in this chapter--that small farms will be adversely affected by BST. Yet, there may not be a contradiction. It may be true that BST will adversely affect the profitability of smaller farms more than larger farms, but that does not necessarily mean the demise of these smaller dairy farms. They are a resilient group. Their owners survive because they are willing to accept a lower return and still remain in dairying. Many have no viable alternatives, or if they do, are willing to relinquish larger incomes they could earn elsewhere because they wish to remain in dairy production. This role of human capital in determining size and growth among dairy farms is measured and discussed by Sumner and Leiby (1987).

Summary

In summary, we do not know to what extent small dairy farms will be affected by the introduction of BST. It is ironic that we may not even know the answer with certainty after BST becomes available, because of the complex interactions that occur in the dairy industry and the myriad of factors affecting farm size and production. Yet, to argue that BST will have no differential impact by farm size is tenuous at best. The issue is the extent of the impact. It is also rather dubious to state that BST will have no negative consumer health effects and then to also argue that BST will have no negative impact on the economic health of small dairy producers. The scientific evidence may support the statement that BST-produced milk is safe to drink, but the scientific evidence does not support the claim there will be no differential effect by size of herd. In fact, insisting that there will be no effect on dairy farm size makes the public skeptical about the truth of "no negative effect on consumer health."

Acknowledgments

The comments of George Casler and Bernard Stanton are appreciated. Conclusions are the author's.

References

Brander, L., and M. Straus. 1959. "Congruence versus profitability in the diffusion of hybrid sorghums." *Rural Sociology* 24:381-383.

Carley, D., S. M. Fletcher, and D. G. S. Alexander. 1989. *The Adoption of Developing Technologies by Georgia Dairy Farmers*. Research Bulletin 382. Athens, GA: University of Georgia Agricultural Experiment Station.

Chavas, Jean-Paul, and G. Magand. 1988. "A dynamic analysis of the size distribution of firms: The case of the U.S. dairy industry." *Agribusiness* 4:315-329.

Cochrane, W. W. 1958. *Farm Prices: Myth and Reality*. Minneapolis, MN: University of Minnesota Press.

Griliches, Z. 1960. "Congruence versus profitability: a false dichotomy." *Rural Sociology* 25:354-356.

_____. 1962. "Profitability versus interaction: another false dichotomy." *Rural Sociology*. 27:327-330.

Hallberg, M. C., and R. Parsons. 1986. "Who Will Gain and Who Will Lose from Production Technologies in the Dairy Industry?" Staff Paper 104. University Park, PA: Pennsylvania State University Department of Agricultural Economics and Rural Sociology.

Havens, A. E., and E. M. Rogers. 1961. "Adoption of hybrid corn: profitability and the interaction effect." *Rural Sociology* 26:409-414.

Kislev, Y., and W. Peterson. 1982. "Process, technology, and farm size." *Journal of Political Economy* 92:578-595.

Lesser, W., W. Magrath, and R. Kalter. 1986. "Projecting adoption rates: application of an *ex ante* procedure to biotechnology products. *North Central Journal of Agricultural Economics* 8:159-174.

Magrath, W. B., and L. W. Tauer. 1988. "New York milk supply with bovine growth hormone." *North Central Journal of Agricultural Economics* 10:233-241.

Marion, B. W., and R. L. Wills. 1990. "A prospective assessment of the impacts of bovine somatotropin: a case study of Wisconsin." *American Journal of Agricultural Economics* 72:326-336.

Marsh, W. E., D. T. Galligan, and W. Chalupa. 1988. "Economics of recombinant bovine somatotropin use in individual dairy herds." *Journal of Dairy Science* 71:2944-2958.

Office of Technology Assessment. 1986. "Technology, Public Policy and the Changing Structure of Agriculture." OTA-F-285, Washington, D.C: U.S. Congress.

Schmidt, G. H. 1989. "Economics of using bovine somatotropin in dairy cows and potential impacts on the U.S. dairy industry." *Journal of Dairy Science* 72:737-744.

Stanton, B., and D. Bertelsen. 1989. "Operating Results for Dairy Farms Classified by Size, FCRS Data, United States, 1987." Department of Agricultural Economics, A.E. Res. 89-23. Ithaca, NY: Cornell University.

Sumner, D. A., and J. D. Leiby. 1987. "An econometric analysis of the effects of human capital on size and growth among dairy farms." *American Journal of Agricultural Economics* 69:465-470.

Tauer, L. W., and H. M. Kaiser. 1991. "Optimal dairy policy with bovine somatotropin." *Review of Agricultural Economics* 13:1-17.

Weersink, A., and L. W. Tauer. 1990. "Regional and temporal impacts of technical change in the U.S. dairy sector." *American Journal of Agricultural Economics* 72:923-934.

____. (1991 forthcoming). "Causality between dairy farm size and productivity." *American Journal of Agricultural Economics* 73:

Yonkers, R. D., J. W. Richardson, R. D. Knutson, and B. B. Buxton. 1986. "Accomplishing Adjustment in the Dairy Industry During Technological Change: The Case of Bovine Growth Hormone." Texas Agricultural Experiment Station TA-21797, College Station, TX: Texas A&M University.

Zepeda, L. 1990. "Predicting bovine somatotropin use by California dairy farmers." *Western Journal of Agricultural Economics* 15:55-62.

PART FOUR

Bovine Somatotropin and the Market Place

10

BST and the Price of Milk and Dairy Products

R. F. Fallert and M. C. Hallberg

History has shown that the principal beneficiary of technological advance in agriculture is the consumer. Several major technological advances have occurred to sustain agricultural productivity growth of an average of 2.2 percent per year since 1947. Productivity in the food processing sector has grown at a slower rate; an average of 0.7 percent per year since 1949 (Reilly 1989). Labor productivity in food retailing has declined in the 1980s. Over the next 20 years, biotechnology promises to be one of the chief productivity enhancers in milk production.

Few issues have generated as much debate in the popular press and elsewhere as has the potential adoption of recombinant or synthetically produced bovine somatotropin (BST) by the nation's dairy farmers. Some oppose the adoption of BST because it threatens to lead to increased milk production, putting added pressure on the federal government already holding surplus dairy products. Others maintain that adoption of BST over the long-term will lead to the demise of at least some of the smaller dairy farms as milk prices and, therefore, unit returns to dairy farmers decline. Some question whether the marketing sector will pass the expected farm-level milk-price reductions on to consumers. A few are fearful that milk from cows administered BST will be unsafe for human consumption.

Several of these issues are treated at length in other chapters of this volume. The present chapter provides an assessment of the expected impacts of BST adoption on regional milk prices, consumption, and production in the United States, and examines the likelihood that a reduction in farm prices of milk, as a consequence of BST adoption, will be passed on to consumers. To accomplish the first of these tasks, we estimate the long-run regional price and quantity consequences of various assumptions about BST use in the United States. These estimates are made with the aid of a spatial equilibrium model constructed to reflect the basic aspects of existing dairy industry structure and policy. The second objective will be met by examining past pricing behavior of the marketing sector, and specifically the rate and extent to which farm-level price changes are transmitted to the consumer.

General Impacts of Bovine Somatotropin

Animal scientists have shown that BST administration will under certain conditions increase milk yields by as much as 30 to 40 percent (Chalupa 1986), although more realistic projections suggest aggregate milk yield increases will probably be closer to 10 percent (Fallert, et al. 1987). Similarly, animal scientists project increased nutrient intake for cows administered BST (Muller, Chapter 3). Thus, feed costs will increase with BST adoption, but economists, incorporating additional assumptions about added costs for BST and its administration and added animal health costs, have shown that dairy farmers will realize sizable monetary gains from BST adoption at current milk prices and feed costs (Kalter et al. 1984, Milligan 1985, Fallert et al. 1987, Yonkers et al. 1989, and Greaser, Chapter 8).

Bovine somatotropin is a size-neutral technology in the sense that farmers with small herds will benefit as much on a *per-cow basis* as will farmers with larger herds so long as managerial ability is equal. This, then, leads to the general conclusion that all farmers will indeed adopt this technology in their search for greater unit profits. As has been true in the past, innovative farmers will be the first to adopt BST. As other farmers find that their neighbors' profit levels rise, they too will adopt this new technology, albeit with some delay, as they strive to increase profits.

In the longer term, when virtually all dairy farmers will have adopted BST, market prices will fall because consumers will be unwilling to purchase, at current prices, the increased quantity of milk placed on the market. (Of course, market prices are dependent upon government support prices which, in turn, are dependent upon the decisions of policymakers. We expect support prices to fall as well, as policymakers attempt to discourage overproduction of milk, and thus seek to ease the government burden associated with purchasing surplus dairy products.) Because of the reduced milk prices, at least some of the smaller dairy farmers can expect to be squeezed out of business because they will not be able to produce the volume necessary to provide a standard of living satisfactory to the farm family. This is an example of the "technology treadmill" made popular by Cochrane (1958).

Conceptually, the long-run impacts of adoption of BST-like technology can be viewed with the aid of Figure 10.1. Here the curve labeled D represents the industry demand schedule and shows how consumers can be expected to respond to alternative prices. The curve labeled S represents the industry supply schedule and shows how producers can be expected to respond to alternative prices. The point at which these two curves intersect identifies the equilibrium price, P, and the equilibrium quantity, Q.

The curve labeled S' in Figure 10.1 suggests how the supply schedule can be expected to shift after 100 percent of the producers have adopted the BST-like technology. The position of this new supply curve reflects the fact that at any given level of industry production, say Q, farmers are still willing to produce that quantity even if price falls below P because the new technology enables them to produce at lower cost. The new long-run equilibrium price and quantity (P' and Q', respectively) are given by the intersection of this new supply curve with the demand curve.

In Figure 10.1 we see that with adoption of the new technology, equilibrium price falls while equilibrium quantity increases. Consumers in the aggregate will gain because they can buy the commodity for less money than they could prior to adoption of the output-increasing technology. Note, however, that price falls by a greater percentage than quantity increases. Thus aggregate producer revenue falls. This happens because consumers do not respond greatly to price changes. That is, when prices rise, consumers reduce their consumption by a relatively smaller percentage than the percentage by which prices rise. Similarly, when prices fall, consumers increase their consumption by a relatively smaller percentage than the percentage by which prices fall. Such consumer reaction is common to most agricultural industries, including the U.S. dairy industry.

Clearly the U.S. dairy industry is much more complex than implied in Figure 10.1, but the diagram is useful in helping us to understand what will happen in this industry if BST is adopted. Indeed, in the analysis to follow, we determine the consequences of BST adoption in precisely this manner. In general terms, then, we can already

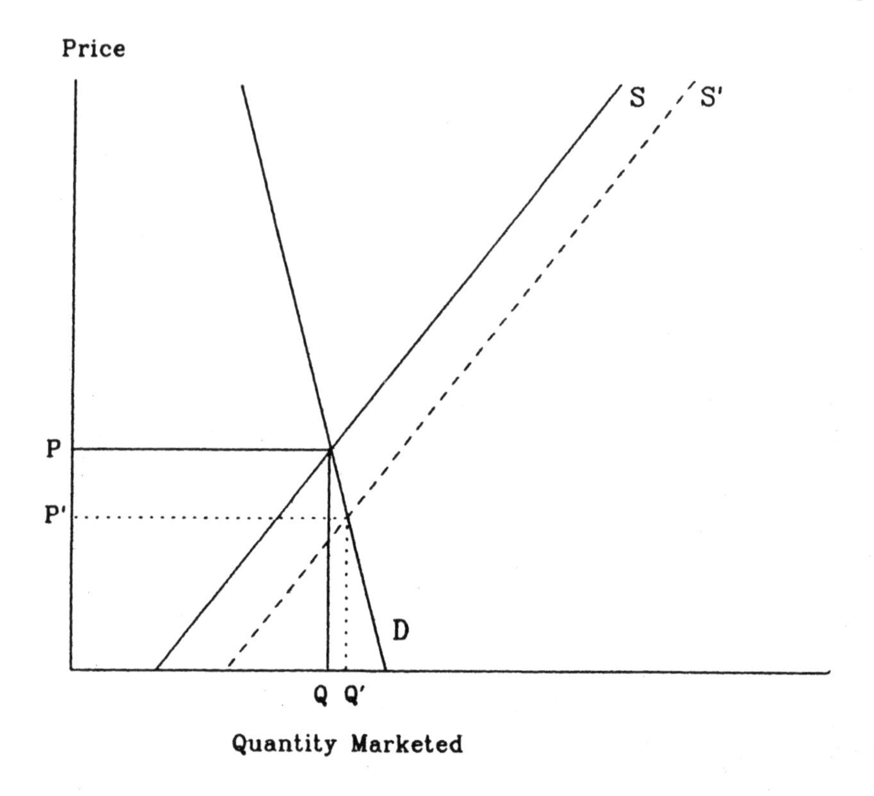

FIGURE 10.1. Long-run consequences of adoption of output-increasing technology.

anticipate the major consequences of BST adoption: in the long-run, milk prices will surely fall and milk production will surely increase, and milk prices will fall by a greater percentage than will milk production increase. In the end, consumers will be better off than they were prior to BST adoption, but farmers in the aggregate may be less well off even though in the short-run (i.e., before all farmers have adopted BST and before milk prices reach their final equilibrium level) early adopters of BST may reap substantial benefits.

The Model

Model Overview

The empirical model used in this analysis is described in Hallberg, et al. (1978). It is a static, spatial, partial-equilibrium model structured to reflect the essential features of existing private and public institutional arrangements governing the U.S. dairy industry. It is designed to generate long-run equilibrium (market-clearing) *farm-level* prices and quantities for each of several regions.

The model is static in the sense that it does not simulate the adjustment path of the dairy industry over time. It merely projects an equilibrium sometime into the future without indicating the year in which equilibrium will occur. It is spatial in the sense that it takes into account interregional interactions and yields a geographic pattern of regional prices and production levels. It is a partial-equilibrium model in that it does not consider interrelations or feedback loops between the dairy sector and other sectors of the agricultural and nonagricultural economies.

Other assumptions made and constraints observed in constructing and operating the model include:

1. Producer responses in any one region can adequately be reflected by a single supply function unique to that region.
2. Government purchases of dairy products are assumed to be known and equal to the equivalent of about 2 percent of total U.S. milk production in any year.
3. Classified pricing is assumed to be maintained under the scenarios examined with this model although here only two use-classes of milk are defined (fluid milk and manufacturing milk) and the price differential between these two use-classes is allowed to be much lower than under existing federal marketing order pricing policy.
4. The producer price of milk is a weighted average of the class prices, with weights equal to the respective proportions of milk used in the two use-classes.
5. The per-hundredweight prices of milk used in a given use-class in any two regions must not differ by more than the cost of transporting an equivalent amount of milk between the respective regions.

About 9 percent of the milk produced in the United States is grade B milk sold to processing plants *not* regulated by federal or state marketing orders. Grade B milk is produced on farms that do not meet the sanitary standards required for producing grade A (fluid) milk; though it can be safely used in most manufactured dairy products.

The remainder of the milk produced in the United States is grade A milk sold to handlers regulated under federal or state marketing orders. The first of the above assumptions, then, is somewhat at variance with the real world. We believe, however, that the supply responses of grade B milk producers do not differ greatly from those of grade A milk producers so that the aggregate supply curve used here is not significantly distorted.

In reality, the Commodity Credit Corporation is obligated by law to purchase selected manufactured dairy products on the open market at prices set so as to support the producer price of raw milk at a legislatively determined level. In recent years the level of Commodity Credit Corporation purchases of dairy products on a milk equivalent basis has been as high as 12 percent of total U.S. milk production. However, our aim in this chapter is to estimate market clearing (equilibrium) levels of prices and quantities--not to determine what support prices will or should be, nor to project the level of government purchases of dairy products, nor even to condone or condemn existing price support policy. Nevertheless, even if the milk price-support program was to be abandoned, the federal government would still be required to take some dairy products off the market in order to meet the needs of its mandated food-aid programs. Thus, assumption 2 was maintained in order to add an element of realism to the analysis. The level of government purchases assumed here was intended to reflect the minimum level of government purchases needed, determined arbitrarily on an examination of the ratio of Commodity Credit Corporation removals to total production over the past several decades and not on an assessment of actual program needs. As this level of government purchases was maintained for all BST scenarios examined, it will have little if any effect on our final conclusions.

In the absence of institutional or policy constraints to the contrary, we could expect the differential between farm-level fluid and manufacturing milk prices to be zero, at least in those regions producing substantial amounts of milk in excess of their fluid milk needs. But milk used to produce fluid products must meet higher sanitary requirements than does milk used to produce manufactured dairy products. Recent estimates suggest that it costs farmers 15 cents per hundredweight to produce milk that meets these higher sanitary requirements (U.S. General Accounting Office 1988). Thus the model implemented here was constructed in such a way that the fluid-manufacturing price differential can be no lower than 15 cents per hundredweight in any one region. This constraint allows the model to find a solution for the entire industry that maximizes industry efficiency.

This assumption is at variance with existing regulatory policy. In the real world, *all* milk regulated by federal marketing orders must meet the higher sanitary requirements, even though more than one-half of it is used in producing manufactured products. Further, the price paid to farmers for fluid milk is sufficiently high to cover the increased costs of meeting the higher sanitary requirements. In the analysis that follows, our aim was to develop equilibrium solutions that represent undistorted, competitive solutions. Again, since our assumption concerning the fluid-manufacturing price differential was maintained throughout all scenarios, it should have little or no effect on our conclusions.

Assumption 4 merely implements a pooling procedure by which average producer (blend) prices are determined. Assumption 5 prevents region y from purchasing milk from region x and then selling it to region z at a profit--something we would not expect to occur in a perfectly competitive environment.

Model Specification

Regional Aggregations. The regional aggregations used in this analysis are indicated in Table 10.1. Basic data were collected for several of the major metropolitan areas of the United States in an effort to maintain as much disaggregation in the basic data set as reasonably possible. Most of these metropolitan areas, however, were aggregated into a subregion, as in Table 10.1. New York City and Chicago were maintained as unique consumption-only subregions; these two cities border on two or more other subregions making further aggregation of these large consumption centers unacceptably arbitrary. In this chapter, results for eight aggregations of subregions are presented, although solutions were obtained for each of the 16 subregions listed in Table 10.1.

Regional Prices. The base data (Tables 10.2 and 10.3) for parameterizing the model were representative of the year 1989. Prices used for fluid milk and for manufacturing milk are representative of 1989 Federal Milk Marketing Order Class I and Class II prices,

TABLE 10.1. Regional Aggregations Used in the Spatial Equilibrium Model.

Region	*Subregion*	*States/Cities Included in Subregion*
Northeast	New England	Maine, New Hampshire, Vermont, Connecticut, Massachusetts, Rhode Island, Boston
	New York	New York (excluding NY City)
	New York City	New York City
	Pennsylvania	Pennsylvania, New Jersey, Philadelphia, Pittsburgh
Southeast	Mid Atlantic	Delaware, Maryland, Virginia, West Virginia, North Carolina, Washington D.C., Baltimore
	East Coast	South Carolina, Georgia, Florida
	Gulf States	Alabama, Mississippi, Louisiana, Arkansas, Tennessee
Midwest	Corn Belt	Missouri, Indiana, Illinois, Kentucky, Iowa, St. Louis
	Chicago	Chicago
	Lake States	Ohio, Michigan, Detroit, Cleveland
Upper Midwest	Upper Midwest	Wisconsin, Minnesota, Minneapolis, St. Paul
Upper Plains	Northern Plains	North Dakota, South Dakota, Wyoming, Montana
	Central Plains	Utah, Colorado, Kansas, Nebraska
So. Plains	Southern Plains	New Mexico, Oklahoma, Texas, Houston
West Coast	West Coast	California, Arizona, Nevada, Los Angeles, San Francisco
Northwest	Northwest	Oregon, Washington, Idaho

respectively. They were estimated from Federal Order Market statistics (U. S. Department of Agriculture 1990). Producer (blend) prices were also taken from the same source.

Supply and Demand Functions. The model requires one or more of three behavioral relations for each of the subregions indicated above: a fluid milk demand function, a manufacturing milk demand function, and/or a raw milk supply function. These functions were determined on the basis of estimates of (1) 1989 quantities demanded or supplied, (2) 1989 prices, and (3) demand and supply elasticities.

The 1989 quantities of fluid milk and of manufacturing milk demanded were determined by multiplying estimated per-capita consumption by total subregion population and calibrated to set total consumption over all subregions equal to total U.S. fluid milk use and total U.S. manufacturing milk use, respectively, as reported by U. S. Department of Agriculture (1990a). Per-capita consumption of fluid and manufacturing milk at the farm level were estimated as explained in Hallberg et al. (1978). Total population and estimates of per-

TABLE 10.2. Basic Data for Spatial Model.[a]

Subregion	*Population (thousand)*	*Per-Capita Consumption* *Fluid*	*Manuf.*
		(pounds)	
New England	12,092	228.8	346.9
New York	6,510	220.5	334.3
New York City	18,004	245.0	371.4
Pennsylvania	14,166	234.8	355.9
Mid Atlantic	23,456	222.4	337.2
East Coast	22,619	200.5	303.9
Gulf States	18,467	200.5	303.9
Corn Belt	11,467	245.0	371.4
Chicago	8,190	245.0	371.4
Lake States	29,500	239.4	362.9
Upper Midwest	9,220	245.0	371.4
Northern Plains	2,656	250.8	380.2
Central Plains	9,148	245.0	371.4
Southern Plains	21,743	211.4	320.5
West Coast	33,730	245.0	371.4
Northwest	8,595	245.0	371.4

[a]Elasticity of demand assumed to be -0.15 for fluid milk and -0.35 for manufacturing milk in all subregions.

TABLE 10.3. Basic Production and Price Data for Spatial Model.

	Quantity Supplied (cwt)	*Supply Elasticity*	*Prices ($/cwt)*		
			Fluid	*Manuf.*	*Blend*
New England	42,482	0.68	14.11	11.03	13.02
New York	111,716	0.68	13.80	11.03	13.00
New York City	--	--	13.96	11.03	--
Pennsylvania	103,895	0.68	13.52	11.03	13.05
Mid Atlantic	53,221	1.02	13.88	11.03	13.55
East Coast	42,242	1.38	14.44	11.03	14.42
Gulf States	52,399	1.22	13.98	11.03	13.45
Corn Belt	99,463	1.00	12.77	11.03	12.03
Chicago	--	--	12.26	11.03	--
Lake States	142,407	0.90	12.82	11.03	12.14
Upper Midwest	341,985	0.85	12.17	11.03	11.88
Northern Plains	32,185	0.46	12.29	11.03	11.37
Central Plains	50,714	1.00	13.10	11.03	12.08
Southern Plains	76,653	1.32	14.00	11.03	13.25
West Coast	212,242	1.35	13.12	11.03	11.11
Northwest	82,970	1.26	12.71	11.03	11.75

capita consumption for each subregion are shown in Tables 10.2 and 10.3. Demand elasticities used (see footnote, Table 10.2) were chosen on the basis of a review of recent research on demand relations for the U.S. dairy industry. These elasticities were assumed to be the same for all subregions included in the model and are intended to represent long-run consumer responses. Subregion supply elasticities were based on the recent work of Chavas and Klemme (1986). Here again we chose elasticities to represent long-run producer responses. Our supplyelasticities range from 0.68 to 1.38 which are in the magnitude of ten-year responses reported by Chavas and Klemme.

The government sector was included as a special region with no supply function, no fluid milk demand function, but a manufacturing-milk demand function that was perfectly vertical at the level of 2.9 billion pounds of milk equivalent--approximately 2 percent of U.S. milk production in 1989.

Total milk production in each subregion was obtained from National Agricultural Statistics Service (U. S. Department of Agriculture 1990).

Costs of Transportation

Unit costs of transportation were based on information obtained by personal contact with industry economists knowledgeable about the costs of and technology used in transporting milk and dairy products. The cost of transporting fluid milk in dollars per hundredweight per 100 miles was assumed to be given by the following relationships:

$$
\begin{aligned}
C_f &= 0.2854 + 0.000045D && \text{if } D \leq 225 \text{ miles} \\
C_f &= 0.3518 + 0.000045D && \text{if } 225 < D \leq 450 \text{ miles} \\
C_f &= 0.4288 + 0.000045D && \text{if } 450 < D \leq 675 \text{ miles} \\
C_f &= 0.4594 + 0.000045D && \text{if } D > 675 \text{ miles,}
\end{aligned}
$$

where D is distance in miles. The cost of transporting milk in manufactured product form in dollars per hundredweight per 100 miles was assumed to be given by the following relationship:

$$C_m = 0.07815 + 0.0000043D.$$

Miles between subregions were determined by (1) locating the longitude-latitude coordinates of the center of a city nearest the midpoint of each subregion, and (2) calculating straight-line distances (in miles) between the points so located.

Interpreting Model Results

The model constructed here is not intended to produce projections of what farm level milk prices and quantities will actually be if BST is adopted. The model cannot do this because it does not consider other factors likely to change over time (feed, labor, and other costs of producing milk, per-cow milk yields due to technologies other than BST or due to genetic improvements, consumer tastes and preferences for dairy products, disposable income, foreign trade in dairy products, etc.) Further, as discussed in the previous section, assumptions incorporated in the model are at variance with existing government policy, and we cannot predict, with certainty, future policy for this industry. For these reasons a solution is obtained for a base situation in which all of the assumptions of the previous section are incorporated, but in which BST is assumed *not* to be adopted. All other scenarios examined assume BST is adopted, shifting the subregion supply curves as, for example, in Figure 10.1. The solutions of the latter scenarios are to be compared to the base situation. Hence, the results generated can be viewed as what might be expected if BST is adopted, and *with all other factors assumed in the base situation remaining unchanged*. In this way we focus on the impact of BST alone, rather than confound the BST impact with the impacts of other factors.[1]

Scenarios Examined

Base Run

This case serves as the *status quo* situation to which all BST-adoption scenarios examined are to be compared. It gives the equilibrium solution assuming BST is *not* adopted (or is prohibited) in all subregions of the United States. As in all subsequent scenarios examined, the base run solution is generated subject to all assumptions as described above.

BST Adopted in All Subregions

In this case we assume BST is adopted in all production subregions of the United States and that the response to BST is identical in all production subregions. The supply curves in each and every production subregion were accordingly shifted to the right by 10 percent as compared to the base situation.

---

[1]For alternative approaches to projections of the consequences of BST adoption when some of these other factors are also permitted to change, see Gum and Martin (1990) and Hallberg (1989).

BST Adopted in All Subregions Except the Upper Midwest

Here we assume that use of BST is prohibited in Minnesota and Wisconsin so that the supply curve in this subregion remains as in the base situation while the supply curves in all other production subregions were shifted to the right by 10 percent.

BST Adopted in All Subregions Except the Northeast

Here we assume that BST use is prohibited in the Northeast subregions while all other subregions adopt this technology. As in the previous case, the supply curves in all subregions of the Northeast remain as in the base situation while the supply curves in all other production subregions were shifted to the right by 10 percent.

BST Adopted and Consumption Reduced in All Subregions

Some research suggests that consumers may react negatively to consumption of milk produced from cows administered BST (Smith and Warland, Chapter 11). To pursue the consequences of this eventuality, the final scenario we examined assumed that BST is adopted throughout the contiguous United States. All subregion supply curves are shifted to the right by 10 percent, but consumers' reaction to the use of BST is assumed to be such that all subregion demand curves for *fluid milk* are shifted to the left by 5 percent as compared to the base situation.

Model Results

BST Adoption in All Subregions

Model results for the 5 scenarios examined are briefly summarized in Table 10.4. Looking first at the consequences of BST adoption in all subregions, compared to the base situation, average producer prices are expected to decline by about 8 percent in all subregions while quantity produced and marketed is expected to increase by only about 2 percent. This is as we anticipated earlier (see Figure 10.1) and follows from the fact that the demand for dairy products is highly inelastic.

Aggregate consumer expenditures on dairy products would decline by 5.7 percent as compared to the base situation while aggregate producer revenue from the sale of milk would fall by 6 percent. The total number of cows required to produce the additional milk, though, would decline by about 7 percent.

Market shares of the different regions would not be affected greatly although the share of the nation's milk supply produced in the Northeast and Upper Midwest would increase slightly while that of the West Coast, Northwest, and Southern Plains would decline slightly. Apparently those regions that now enjoy a slight competitive advantage and are nearest the large population centers will continue to enjoy a competitive edge as prices fall. As milk prices fall relative to costs of transportation, the economic advantages of transporting milk long distances also fall. Consequently, those producers located nearest to consumers should be expected to supply somewhat more of their local fluid milk needs. This clearly works to the advantage of producers located near the large population centers in the eastern United States.

BST Ban in Selected Regions

If BST were prohibited in the Upper Midwest region, but adopted elsewhere in the nation, producer prices in the Upper Midwest would still fall commensurate with price declines nationwide, but milk produced and marketed in that region would actually fall by about 5 percent as compared to the base situation. This is due, of course, to the fact that cows are now less productive in the Upper Midwest than elsewhere. The market share of the Upper Midwest would fall from 23.3 percent to 21.8 percent. The Northeast would be the principal gainer from a BST ban in the Upper Midwest as its market share would increase from 17.8 percent to 18.1 percent. The West Coast would also gain some market share, but less than the Northeast.

If, on the other hand, BST were prohibited in the Northeast while all other subregions adopt BST, producer prices in the Northeast would fall commensurate with the fall in producer prices nationwide. Production in the region would fall by about 4 percent as compared to the base situation. In this case, the Upper Midwest would be the

TABLE 10.4. Equilibrium Solutions for BST Scenarios.

Region[a]	Base Run	BST Adopted Everywhere	BST Banned in U. Midwest	BST Banned in Northeast	BST & Reduced Consumption
		Average Producer Prices ($/cwt)			
Northeast	$12.44	$11.42	$11.69	$11.73	$10.92
Southeast	13.61	12.66	12.82	12.77	12.16
Midwest	12.18	11.21	11.48	11.42	10.71
Upper Midwest	11.66	10.67	10.96	10.87	10.19
Upper Plains	12.29	11.33	11.36	11.38	10.81
Southern Plains	13.17	12.22	12.44	12.38	11.73
West Coast	11.84	10.90	11.11	11.05	10.47
Northwest	10.99	10.17	10.31	10.23	9.68
U.S.	12.19	11.23	11.47	11.40	10.74
		Regional Supply/Marketings (mil lbs)[b]			
Northeast	25,507	26,739	27,107	24,516	26,044
	(17.3)	(17.8)	(18.1)	(16.4)	(18.0)
Southeast	14,826	15,089	15,313	15,248	14,432
	(10.1)	(10.0)	(10.2)	(10.2)	(10.0)
Midwest	24,807	25,410	25,919	25,799	24,453
	(16.8)	(16.9)	(17.3)	(17.2)	(16.9)
Upper Midwest	34,296	35,331	32,568	35,830	34,120
	(23.3)	(23.5)	(21.8)	(23.9)	(23.6)
Upper Plains	8,765	9,081	9,070	9,094	8,800
	(6.0)	(6.0)	(6.1)	(6.1)	(6.1)
Southern Plains	7,750	7,796	7,969	7,921	7,414
	(5.3)	(5.2)	(5.3)	(5.3)	(5.1)
West Coast	23,542	23,255	23,812	23,634	22,127
	(16.0)	(15.4)	(15.9)	(15.8)	(15.3)
Northwest	7,773	7,871	8,002	7,926	7,424
	(5.3)	(5.2)	(5.3)	(5.3)	(5.1)
U.S.	147,266	150,572	149,761	149,968	144,815
		Consumer Expenditures and Producer Revenue in the U.S. ($mil)			
Expenditures	$18,362	$17,314	$17,576	$17,523	$15,944
Revenue	$17,951	$16,901	$17,172	$17,101	$15,550
		Number of Cows Required (thousand)[c]			
Number of Cows	10,353	9,622	9,779	9,741	9,255

[a] See Table 10.1 for regional definitions.

[b] Numbers in parentheses represent the respective region's share of the national market.

[c] Number of cows estimated on the basis of 1989 milk production per cow (14,225 lbs) when BST was assumed not to be used, and 110 percent of 1989 milk production per cow (15,648 lbs) when BST was assumed to be used.

principal gainer. The West Coast would benefit less from a BST ban in the Northeast than from a BST ban in the Upper Midwest.

BST Adoption and Reduced Consumption

This is clearly the worst-case scenario from the perspective of producers. Producer prices would fall an average of 12 percent as compared to the base situation, and quantity produced and marketed nationwide would fall by about 1.7 percent. All regions would suffer about the same price decline. The Northeast and Upper Midwest, however, would benefit slightly in terms of increased market share, while the West Coast would lose market share. Milk production and marketings in the Northeast would actually *increase* slightly, whereas in the West Coast region, production and marketings of milk would decrease. The number of cows required in the nation would decrease by 10.3 percent. Thus, if realized, this scenario would require major adjustments on the part of dairy farmers in the United States.

Effects of BST on Retail Prices

To anticipate the likely impacts on consumer prices of productivity gains in the milk production sector resulting from BST use, one must first bear in mind that the farmer's share of the consumer's food dollar varies widely among different food products. The farmer's share is the average percentage farmers get of each dollar the consumer spends in retail grocery stores. Since this share varies widely among different foods, the likely effects on retail prices of productivity gains in production will vary considerably among foods, and even among individual dairy products. For example, in 1990 the farmer's share of the consumer food dollar ranged from 64 percent for eggs, 45 percent for fresh milk, 34 percent for natural cheddar cheese, and 6 percent for bread and corn flakes (Dunham 1991). In general, the more highly processed the product, the smaller the farmer's share. Other factors affecting farm share include shipping distance from farm to the consumer and product perishability.

Because of the variation in farmer's shares among foods, changes in farm level prices will likely have different effects on retail prices. For example, if the farmer's share of the consumer food dollar is about

50 percent as in fresh fluid milk, a 10 percent decrease in the farm price of milk is likely to result in a 5 percent decrease in the retail price of milk if all other things such as marketing costs remain unchanged. In contrast, since the farmer's share of the consumer food dollar is about 35 percent for natural cheese, a 10 percent decrease in the farm price of milk is likely to result in about a 3.5 percent decrease in the retail price of cheese.

The major factors influencing retail food prices are farm prices, costs of processing and distributing foods, and consumer demand. One cannot, therefore, realistically assume that reductions in farm milk prices will automatically be reflected in lower retail milk prices. For example, the costs of processing and marketing beverage milk and manufactured dairy products might be increasing during the period when farm prices are declining. This was the situation in the latter half of the 1980s. However, analysis shows that retail milk prices during this period were lower than they would have been without declining farm milk prices--even though retail prices were not declining in absolute terms.

Another factor that will affect the correspondence between farm and retail milk and dairy product prices is the degree of flexibility in the dairy price support program. If dairy price supports are allowed to decline as milk production costs decline with BST use, consumers are likely to reap the benefits of this cost-reducing technology. But with a more rigid price support program or initiation of a supply management program (stringent milk marketing quotas, for example), consumers are less likely to benefit from the milk production efficiency brought about by BST.

A recent study concluded that the introduction of BST would result in a 4.7 cent reduction in the retail price of a gallon of milk (Yonkers et al. 1989). A special report by Consumers Union (a consumer information, product testing, and activist organization), on the other hand, suggested that there was *no* relationship between changes in farm milk prices and retail milk prices (Hansen 1990), although a review of that analysis indicates that a basic error was made in the series of price comparisons. This is only one of a steady stream of economic analyses of the likely effects of BST on the farmer, the consumer, and the industry that have come forth since the early 1980s (see Fallert and Blayney 1991).

Numerous studies have shown that historically there has been a strong correspondence between farm prices and retail prices with little or no lag (Gardner 1975, Gardner 1991, Hahn and Duewer 1991, Kinnucan and Forker 1987, Miller 1983, Outlaw et al. 1991, Wohlgenant and Haidacher 1989). Preliminary research on the correspondence between changes in farm milk prices and retail milk and dairy product prices by Weimar et al. (1991) and U.S. General Accounting Office (1991) indicates that there is a lag between farm and retail price changes; and that this lag is shorter when farm prices are increasing than when farm prices are decreasing. In other words, farm-to-retail margins tend to become narrower when farm prices are increasing and to become wider when farm prices are decreasing. The widening of farm-to-retail price spreads when farm prices fall (in fact, the stickiness of retail prices) has existed since farm milk prices reached a peak of $16 per hundredweight in December 1989 and declined to $11.70 per hundredweight (i.e., by 27 percent) by December 1990.

The work of Weimar et al. (1991) and Shargots (1991) also shows sub-stantial differences in farm-to-retail fluid milk price margins among regions. For example, retail prices in the North Central and Southern regions appear to have been quite responsive to declines in the Federal Milk Marketing Order Class I Price, while retail fluid milk prices in the western and northeastern regions of the United States have been much less responsive to Federal Milk Marketing Order Class I price declines-especially during late 1990 and early 1991.

Given the highly volatile farm milk prices during 1989 and 1990, it appears that the record farm-to-retail price spreads in recent months are an abberation rather than what can be expected in the future based on historic patterns. Hahn and Duewer (1991) also found this same characteristic to be essentially true of farm-to-retail price spreads for pork, and Lambert (1991) for beef.

Summary

Adoption of BST by the nation's dairy farmers will in the long-term lead to a significant drop in farm-level milk prices and to a slight increase in quantity of milk marketed. We estimate that if BST is adopted in all regions of the United States, equilibrium farm-level prices will in the long-run fall by an average of 8 percent, overall marketings of milk will in the long-run increase by only about 2 percent, and the national dairy herd will be reduced by 7 percent.

There will be some variation in marketings by regions, due primarily to the different regional fluid milk requirements. Marketings in the Northeast will increase somewhat more than 2 percent; while marketings in the Southern Plains will change little and those in the West Coast region will actually decrease slightly.

Should any one of the nation's major milk-producing regions prohibit BST use while all other regions adopt this technology, producers that cannot use BST will be disadvantaged because their region's production share of the national market will decline significantly. In addition, producers in the region of the ban would be subject to a price decline commensurate with the decline in prices nationwide as most other producers adopt the new technology. The situation will be worse for producers in regions with low fluid consumption relative to total regional consumption, since it is more expensive to import fluid milk than it is to import manufactured dairy products.

Should the expansion in supply caused by BST adoption be accompanied by a drop in consumer demand, producer prices would fall another 4 percent and milk marketings would fall by about 2 percent. Consumer demand might be expected to drop if a significant number of consumers conclude that milk produced from BST-injected cows is unsafe for human consumption. This situation would require major adjustments on the part of the nation's dairy farmers, because of the potential 10 percent reduction in the size of the national dairy herd.

The impact of BST adoption on retail prices is less clear. Research results on the transmission of farm-level price changes to the retail level are mixed. Research based on several years of price trends suggests that reductions in farm-level prices will be reflected at the retail level with little or no delay. More recent research based on data for the past two or three years suggests that a significant time lag is involved here, and that the lag is shorter when farm prices are increasing than when decreasing. The explanation for this phenomenon appears to be that the 1989-91 period represents an abberation rather than a change in the structure of the dairy marketing-processing sector. Clearly, however, this is a poor excuse for attempting to curtail efforts aimed at decreasing the costs of milk production as, for example, by BST use.

Acknowledgments

We appreciate the helpful comments of Don Blayney, Denis Dunham, Alden Manchester, and Mark Weimer, all of Economic Research Service, U.S. Department of Agriculture, on an earlier draft of this chapter.

References

Blayney, D. P., and R. F. Fallert. 1990. "Biotechnology and agriculture--emergence of bovine somatotropin (BST)." Staff Report AGES-9037. Washington D.C: Economic Research Service, USDA.

Chalupa, W. 1986. "Long-term Responses of Lactating Cows to Daily Injections of Recombinant Somatotropin." Paper presented at American Dairy Science Association annual meetings, Davis CA.

Chavas, Jean-Paul, and R. M. Klemme. 1986. "Aggregate milk supply response and investment behavior on U. S. dairy farms." *American Journal of Agricultural Economics* 68:1:55-66.

Cochrane, W. W. 1958. *Farm Prices: Myth and Reality.* Minneapolis, MN: University of Minnesota Press.

Dunham, Denis. 1991. *Food Cost Review, 1990.* AER-651. Washington D.C: Economic Research Service, USDA.

Fallert, R. F., and D. P. Blayney. 1991. "The economics of BST: impacts on the farmer, the consumer, and the industry." Paper presented at the Annual Meeting of the American Association for the Advancement of Science. Washington D.C.

Fallert, R. F., T. McGuckin, C. Betts, and G. Bruner. 1987. *BST and the Dairy Industry: A National, Regional, and Farm-Level Analysis.* AER-579. Washington D.C: Economic Research Service, USDA.

Gardner, B. L. 1975. "The farm-retail price spread in a competitive food industry." *American Journal of Agricultural Economics* 57:399-409.

Gardner, B. L. 1991. "Prices and marketing relationships in the U.S. dairy industry." Statement presented to the Subcommittee on Livestock, Dairy, and Poultry Committee on Agriculture, U.S. House of Representatives. Washington D.C.

Gum, R. L., and W. E. Martin. 1990. *Economic Impacts of Biotechnical Innovations in the U.S. and Arizona Dairy and Cotton Industries.* Bulletin No. 267. Tucson, AR: Arizona State University Agricultural Experiment Station.

Hallberg, M. C. 1989. "How far will milk prices fall?" *Choices.* Fourth Quarter.

Hallberg, M. C., D. E. Hahn, R. W. Stammer, G. J. Elterich, and C. L. Fife. 1978. *Impact of Alternative Federal Milk Marketing Order Pricing Policies on the United States Dairy Industry.* Bulletin 818. University Park, PA: Pennsylvania State University Agricultural Experiment Station.

Hahn, W. F., and L. A. Duewer. 1991. "An evaluation of farm to retail price spreads for pork in 1990." Paper presented at the Annual Meetings of the American Agricultural Economics Association, Manhattan, Kansas.

Hansen, Michael K. 1990. "Biotechnology and milk: benefit or threat? An analysis of issues related to BGH/BST use in the dairy industry." Mount Vernon, NY: Consumer Policy Institute, a Division of Consumers Union.

Kalter, R. J., A. Milligan, W. Lesser, W. Magrath, and D. Bauman. 1984. "Biotechnology and the Dairy Industry: Production Costs and Commercial Potential of the Bovine Growth Hormone." Research Report 84-22. Ithaca, NY: Cornell University Department of Agricultural Economics.

Kinnucan, H. W., and O. D. Forker. 1987. "Asymmetry in farm-retail price transmission of major dairy products." *American Journal of Agricultural Economics* 69:2885-2.

Lambert, Chuck. 1991. "Retail margins: Are they really that wide?" *Beef Business Bulletin*. National Cattleman's Association. Vol. 14, No. 36.

Miller, J. J. 1983. "A longrun look at margins and productivity in fluid milk processing." *Dairy Outlook and Situation*. DS-394. Washington D.C: Economic Research Service, USDA.

Milligan, R. A. 1985. "The Economic Implications of Bovine Growth Hormone on Dairy Farm Profitability." Staff Paper No. 85-16. Ithaca, NY: Cornell University Department of Agricultural Economics.

Office of Technology Assessment. 1991. "U.S. dairy industry at a crossroad: biotechnology and policy choices." Special Report, OTA-F-470. Washington D.C: U.S. Government Printing Office.

Outlaw, J., O. Capps, Jr., R. Knutson, and R. Schwart, Jr. 1991. "The correspondence of farm and retail milk price movements." AFPC Policy Working Paper 91-7. College Station, TX: Texas A&M University Department of Agricultural Economics.

Reilly, J. M. 1989. "Consumer effects of biotechnology." AIB-581. Washington D.C: Economic Research Service, USDA.

Shargots, R. 1991. "An assessment of farm to retail price spreads for milk and selected dairy products." Working paper for U.S. Congress. Washington D.C: General Accounting Office.

U. S. Department of Agriculture. 1990. *Federal Milk Order Market Statistics, 1989 Annual Summary*. Statistical Bulletin No. SB-810. Washington D.C: Agricultural Marketing Service.

U. S. Department of Agriculture. 1990a. *Milk Production, Disposition and Income, 1989 Summary*. Washington D.C: National Agricultural Statistics Service.

U. S. General Accounting Office. 1988. "Milk Marketing Orders: Options for Change." GAO/RCED-88-9. Washington, D.C.

U. S. General Accounting Office. 1991. "Federal Dairy Programs: Information on Farm and Retail Milk Prices." GAO/RCED-91-187FS. Washington, D. C.

Weimar, M., W. Haha, and J. Mengel. 1991. "An analysis of aggregate and regional retail price spreads for milk and selected dairy products." (forthcoming). Washington D.C: Economic Research Service, USDA.

Wohlgenant, M. K., and R. C. Haidacher. 1989. "Retail to farm linkage for a complete demand system of food commodities." Technical Bulletin 1775. Washington, D.C: Economic Research Service, USDA.

Yonkers, R. D., J. W. Richardson, and R. D. Knutson. 1989. "Regional Farm-Level Impacts of BST." Agricultural and Food Policy Center Working Paper 89-7. College Station, TX: Texas A&M University Department of Agricultural Economics.

11

Consumer Responses to Milk from BST-Supplemented Cows

Blair J. Smith and Rex H. Warland

If the commercial use of bovine somatotropin (BST) is finally approved by the Food and Drug Administration, it will be adopted by dairy farmers initially on the basis of their perceptions of its economic benefits. Once the milk from BST-treated cows (if any) reaches the market, consumers will decide for or against such milk on the basis of their own notions of its intrinsic qualities.

Consumer demand for milk is generally thought to be inelastic. That is, when the price of milk goes up they don't buy much less, and when the price of milk goes down they don't buy much more. The other side of this phenomenon of milk-price elasticity is that relatively small changes in purchases of milk result in relatively large changes in the price of milk, again in opposite directions.

Few industry observers, if any, expect milk consumption to increase *because* it is produced by cows that have been administered BST. It seems unlikely, however, that farmers will adopt BST if it does not reduce their cost of producing milk. The structure of the industry is such that competition would be expected to eventually force production economies on to the consumer. The magnitude of such economies will depend quite heavily on the price farmers must pay for BST, yet to be established.

Even if one can logically anticipate an increase in milk consumption because of reduced farm costs and lower consumer prices, the increase is likely to be quite small. It could pale in comparison to the negative effect that consumer fears of milk from BST-treated cows might have on the consumption of milk. If, indeed, the net effect of positive and negative forces impacting on milk consumption does bring about reduced consumption, dairy farmers, milk dealers, and milk consumers alike stand to lose. Farmers will lose because less milk will be sold and at lower prices, dealers will lose because of reduced plant through-put, and consumers will lose because they will have to replace the nutrients now derived from milk with nutrients from less-preferred and/or higher-priced foodstuffs.

The aim of this chapter is to report the findings of the several state and national studies that have attempted to assess consumer concerns and their possible reaction to milk from BST-supplemented cows. With such information, the industry can better gauge the magnitude of any effort that might be needed to deal with the BST issue, and is likely to have better insight into the type of approach that would be most effective.

Synopses of the Major Studies

The earliest known attempt to obtain insight into consumer perceptions, possible concerns, and potential reactions to the use of BST in the United States consisted of six focus-group interviews conducted for the National Dairy Board by Fine, Travis, and Associates, Inc. (1986).

Each of the six focus groups was composed of 8 to 10 women, all with children 17 years of age or younger living at home. There were two such groups in each of three cities (New Rochelle, NY, St. Louis, MO, and Concord, CA).

Focus-group research is not ordinarily representative of any general population, so the findings of those interviews will not be reported here. Nevertheless, scholars conducting the more-representative studies reported below did benefit from the focus-group research in those cases where the results were available.

At this time, we know of eleven consumer-focused BST studies that involved reasonably large numbers of subjects. Some of these studies dealt primarily with BST issues as a direct point of focus. Others embedded questions about BST in broader contexts of food concerns and choices, or biotechnology in general. The amount of pre- and survey-in-progress conditioning of respondents in regard to BST also varied. Certain key characteristics of the eleven studies are described in Table 11.1.

TABLE 11.1. Key Characteristics of BST-related Consumer Studies.

Study	*Characteristics*
AHI	Conducted nationwide by Center for Communications Dynamics for the Animal Health Institute. February, 1987. **Subjects:** 1,315 adults. **Method of interview:** Phone. **Representativeness of sample:** Unknown. **Primary focus of survey:** Biotechnology in general and BST in particular. **Information about BST provided respondents:** Both pro and con arguments regarding BST were provided. **Tone of survey and BST questions:** Unknown. **Study source:** Animal Health Institute. (date unknown) "Bovine somotatropin (BST)." 119 Oronoco Street, Box 1417-D50. Alexandria, VA 22313.
PSU STATE	Conducted in Pennsylvania by the Department of Agricultural Economics and Rural Sociology, Pennsylvania State University. Spring 1989. **Subjects:** 508 adults. **Method of interview:** Phone. **Representativeness of sample:** Representative. **Primary focus of survey:** Food choice behavior generally. Heavy emphasis on milk. BST questions specific but a minor part of the questionnaire. **Information about BST provided respondents:** "Suppose you learn one day that milk is being produced by cows that are being given a hormone to increase the amount of milk they produce." **Tone of survey and BST questions:** Use of the word hormone, although accurate, may have been threatening or alarming to some respondents. **Study source:** Smith, Blair J. 1991. "Consumer and public school perspectives on the BST issue in Pennsylvania." *Farm Economics*. March/April. Pennsylvania State University, Cooperative Extension Service. University Park, PA 16802.
DAIRY TODAY	Conducted nationwide by Consumer Network, Inc. for *Dairy Today*. July, 1989. **Subjects:** 180 adult food shoppers. **Method of interview:** Mail. **Representativeness of sample:** Not representative. **Primary focus of survey:** Eight of thirteen questions related directly to BST.

(continues)

TABLE 11.1. *(continued)*

Study	*Characteristics*
	Information about BST provided respondents: Called it a protein product that is the outgrowth of biotechnology to stimulate milk production and make dairy farmers more efficient. Later referred to it as a protein hormone. Said it will not influence milk taste and is not expected to affect the price of milk. **Tone of survey and BST questions:** Factual and as neutral as possible. Yet, both the words biotechnology and hormone would be expected to elicit some negative reaction. **Study source:** Henderson, Pam C. 1989. "Consumers balk at BST." *Dairy Today*. November/December. 230 West Washington Square, Philadelphia, PA, 19105. Mr. Jim Dickrell, Editor of *Dairy Today* provided additional information from the study conducted by Consumer Network, Inc., 3624 Market Street, Philadelphia, PA 19104.
VA	Conducted in Virginia by Department of Agricultural Economics, Virginia Polytechnic Institute and State University. Fall, 1989. **Subjects:** 607 primary grocery shoppers. **Method of interview:** Mail. **Representativeness of sample:** Representative. **Primary focus of survey:** BST. **Information about BST provided respondents:** Two different half-page discussions of the pros and cons of BST preceded the questions, one purported to be more neutral than the other. **Tone of survey and BST questions:** The differences in the two presentations seemed too subtle to expect very much difference in responses. The statements of opposition to BST are factual, but would probably induce a negative bias toward BST. **Study source:** Preston, Warren P., Anya M. McGuirk, and Gerald M. Jones. 1991. "Consumer Reaction to the Introduction of Bovine Somatotropin." In Julie A. Caswell (ed). *The Economics of Food Safety*. Chapter 10. New York, NY: Elsevier Science Publishing Company Inc.; and McGuirk, Anya M., Warren P. Preston, and Gerald M. Jones. 1990. "Biotechnology and the consumer: the case of BST." SP-90-6. Virginia Polytechnic Institute and State University, Department of Agricultural Economics. Blacksburg, VA 24061.
WA	Conducted in Washington by Norris and Associates for the Washington State Dairy Products Commission, 1107 N.E. 45th Street, Seattle, WA, 98105. Spring, 1990. **Subjects:** 400 principal adult food shoppers. **Method of interview:** Phone.

(continues)

TABLE 11.1. *(continued)*

Study	*Characteristics*
	Representativeness of sample: Representative. **Primary focus of survey:** BST. **Information about BST provided respondents:** None, initially. Then, BST and BGH were asserted to be the same thing--a growth hormone developed through biotechnology to be injected into cows to encourage more milk output. **Tone of survey and BST questions:** Those initially opposed to BST/BGH were provided with a series of positive messages, those in favor a series of negative messages. **Study source:** Norris and Associates. 1990. "Washington Dairy Products Commission BST Statewide Consumer Survey Wave II." 17111 161-st Avenue, N.E., Woodinville, WA 98072.
NY	Conducted in New York by Department of Agricultural Economics, Cornell University. Spring, 1990. **Subjects:** 716 primary grocery shoppers. **Method of interview:** Mail. **Representativeness of sample:** Representative. **Primary focus of survey:** General concerns about food safety with emphasis on BST and antibiotics in milk. **Information about BST provided respondents:** Similar to that which was provided to respondents in the VA study. **Tone of survey and BST questions:** Only one description of BST was used. It was probably perceived as more neutral than either of the two VA versions. **Study source:** Kaiser, H. M., C. W. Scherer, and C. M. Barbano. 1991. "Consumer perceptions and attitudes towards bovine somatotropin." Paper submitted for publication in *Northeastern Journal of Agricultural and Resource Economics*.
WI	Conducted in Wisconsin by Department of Consumer Science, University of Wisconsin. Spring, 1990. **Subjects:** 1,056 primary adult food shoppers. **Method of interview:** Phone. **Representativeness of sample:** Representative. **Primary focus of survey:** BST.

(continues)

TABLE 11.1. *(continued)*

Study	*Characteristics*
	Information about BST provided respondents: "In recent months there has been a controversy in Wisconsin about dairy farmers potentially using a protein hormone on their cows to increase milk production. You may have seen or heard the hormone referred to by one of two names, either bovine growth hormone, BGH, or bovine somatotropin, BST. Are you aware or not of these recent controversies regarding Wisconsin dairy farmers' use of BGH?" **Tone of survey and BST questions:** Since no attempt to provide information on both sides of the "controversy" was made, the tone of the survey with respect to BST was neutral at best, perhaps tending toward the negative because of the reference to both BGH and BST as protein hormones. **Study source:** Douthitt, Robin A. 1990. "Wisconsin consumer's attitudes towards bovine somatotropin (BST) and dairy product labeling." University of Wisconsin, Department of Consumer Science. Madison, WI 53706.
MO	Conducted in Missouri by Department of Family Economics and Management, University of Missouri. Spring, 1990. **Subjects:** 219 principal food shoppers. **Method of interview:** Mail and phone. **Representativeness of sample:** Representative. **Primary focus of survey:** Biotechnology and food safety in general with a special emphasis on milk and BST. **Information about BST provided respondents:** Positive commentary on biotechnology and genetic-engineering, followed by a series of neutral introductions to questions about BST and milk. **Tone of survey and BST questions:** No direct positive or negative arguments regarding BST are offered, but the use of the term growth hormone may have carried negative connotations. **Study source:** Slusher, Barbara J. 1991. "Consumer acceptance of food production innovations--an empirical focus on biotechnology and BST." In Robert N. Mayer (ed). *Enhancing Consumer Choice*. Pp. 105-116. American Council on Consumer Interests. 240 Stanley, Columbia, MO, 65211.
NDB	Conducted nationwide by The Research Alliance for the National Dairy Board, 2111 Wilson Boulevard, Arlington, VA 22201. Spring, 1990. **Subjects:** 592 primary grocery shoppers. **Method of interview:** Phone-mail-phone. **Representativeness of sample:** Not representative. **Primary focus of survey:** Primary focus on BST with substantial attention to food safety concerns in general.

(continues)

TABLE 11.1 *(concluded)*

Study	*Characteristics*
	Information about BST provided respondents: The pre-exposure telephone survey instrument linked the use of hormones to increased production of milk in dairy cows. **Tone of survey and BST questions:** Generally neutral but the use of the word "hormone" may have raised a red flag in some respondents' minds. **Study source:** The Research Alliance. 1990. "Consumer reactions to the use of BST in dairy cows." Marketing Research and Counseling Services. 901 Battery Street, Suite 202. San Francisco, CA 94111-1301.
PSU SCHOOL	Conducted in Pennsylvania by Department of Agricultural Economics and Rural Sociology, Pennsylvania State University. Spring, 1990. **Subjects:** 429 public school food-service directors. **Method of interview:** Mail. **Representativeness of sample:** Representative. **Primary focus of survey:** Public school milk matters generally with a section (about 20 %) of questionnaire devoted to BST. **Information about BST provided respondents:** "There is talk about someday administering a hormone to dairy cows to stimulate them to produce more milk. This hormone is referred to as bovine somatotropin (BST) or bovine growth hormone (BGH)." **Tone of survey and BST questions:** Same as the PSU-STATE study. **Study source:** Smith, Blair J. 1991. "Consumer and public school perspectives on the BST issue in Pennsylvania." *Farm Economics*. March/April. Pennsylvania State University, Cooperative Extension Service. University Park, PA 16802.
PSU NATION	Conducted nationwide by Department of Agricultural Economics and Rural Sociology, The Pennsylvania State University. Fall, 1990. **Subjects:** 1,200 adults. **Method of interview:** Phone. **Representativeness of sample:** Representative. **Primary focus of survey:** A general study of food choice behavior with an emphasis on frozen and cultured dairy products. Questions on BST were specific, but only a minor part of the overall questionnaire. **Information about BST provided respondents:** None. **Tone of survey and BST questions:** The use of the term "bovine growth hormone" without any explanation of its nature, purpose, or effect probably engendered a somewhat negative attitude toward BST. **Study source:** Unpublished information provided by Rex Warland, Robert Herrmann, and Arthur Sterngold. Pennsylvania State University, Department of Agricultural Economics and Rural Sociology. University Park, PA 16802.

Geographic Scope and Representativeness of the Studies

Four of the eleven studies are national in scope and the remaining seven are confined to persons within the borders of a single state. Only one of the four national studies is deemed to be representative of the United States as a whole. Representativeness is claimed for all seven state studies.

In this chapter, we use a somewhat loose standard for representativeness. For the most part, if the sample was randomly drawn from the population of interest, we regard the results as representative.

The sample used for the AHI study deliberately contained more women than men because, the report said, women are the primary purchasers of dairy products. How the sample was drawn and whether the respondents to the survey might otherwise be representative of the national population is not revealed in materials available to us.

How "national" the *DAIRY TODAY* study is, with only 180 observations, is uncertain. The 180 survey respondents were drawn from a panel of about 5,000 food shoppers, cross-sections of whom are asked each month for their opinions on selections from a pool of more than 17,000 supermarket items. This panel of "shoppers" is not considered to be representative of U.S. shoppers generally, as its members are thought to be better informed on food matters and more highly motivated to be "good shoppers."

Respondents in the NDB study were the principal grocery shoppers in households that are members of the National Family Opinion. Thus, it is not appropriate to suggest that findings of the NDB study are representative of principal grocery shoppers throughout the country.

Representativeness was achieved in all three Penn State studies. The state and national consumer surveys were conducted so as to assure sampling error would be less than 5 percent. The entire population of Pennsylvania's 497 public school food-service directors received questionnaires, and 429 returned useable responses. This response rate (86.3%) assures the representativeness of those findings.

The authors of the Virginia study compared sample statistics to population parameters for certain key socioeconomic variables and found that the survey respondents were older, more highly educated, had higher incomes, and were more likely to be white than the total population. Nevertheless, the initial mailing was made to a random sample of Virginia households, and we feel it is valid to generalize the findings of the study to the Virginia population as a whole.

The authors of the Washington study claim that their sample statistics are within 5 percentage points of the corresponding population parameters.

The Wisconsin researchers purchased a representative sample frame of 2,700 Wisconsin telephone numbers from a commercial vendor. Interviewers asked to speak with the adult person 18 years or older who made most of the household's food purchase decisions. After approximately 1,525 calls 1,056 schedules were completed and serve as the basis for their report.

The workers at the University of Missouri purchased the names and addresses of the 1,200 households used in their original mailing from a commercial vendor. This excluded households with unlisted phone numbers (known to be about 33%, nationally). Thus, although the representativeness of the Missouri study cannot be assured, we are willing to characterize the results as representative of the Missouri populace.

The Cornell University study was based on a stratified random sample of households that deliberately over-sampled the nonmetro, upstate areas to assure adequate representation outside New York City. The respondents, on average, had higher income and education levels and a greater percentage of males than the actual population. The results, however, were adjusted using standard statistical techniques to make them representative of the state as a whole.

Survey Instruments and Procedures

The format of the survey instrument, the wording of its questions, and the background information about BST and related issues given to survey participants could have affected the results of any of the studies listed in Table 11.1. We were able to obtain copies of the survey instruments for each study except that conducted for the Animal Health Institute by the Center for Communications Dynamics, Washington, D.C.

In Table 11.1 we also show the results of our attempt to summarize the instruments and procedures used in the 11 studies, according to the key considerations presented in the paragraph above. Both the tabular data and our comments related to those data are necessarily cryptic at this point; we will more fully examine the results of the studies in the next section of this chapter.

It would be extremely difficult to devise a set of questions about BST that would be perceived to be truly neutral by a majority of survey participants. Such processes or substances as *genetic engineering*, *hormone*, and *injection* are all more likely to contain negative than positive connotations for many people. Even the suggestion that BST would result in the production of more milk probably is not viewed favorably by many people because of the media concern about milk surpluses in recent years. From the perspective of the consumer, the prospect of lower costs and prices may hold little attraction, because milk and dairy products are now a good food bargain. Consumers are not very responsive to changes in the price of milk. Some believe that reductions in cost of production from the use of BST are not likely to be passed along to them.

In view of the negative or anti-BST messages that have already appeared in the media, it may be that the "tone" of most of the questionnaires described in Table 11.1 were about as neutral as they could or needed to be. Once BST is approved for commercial use, we may anticipate an even harsher anti-BST campaign by those opposed to its use on any of several possible grounds. The arguments against BST that have surfaced so far include that it will have a deleterious effect on the cow's health, it will accelerate the loss of dairy farms, and it may have negative consequences for human health of presently unknown import or onset. Thus, although we, as researchers, would have preferred to approach the subject of BST in a purely factual, fully informative, balanced, and neutral sort of way, the topic does not readily lend itself to such treatment within the confines of a mail or telephone survey process. While we might generally characterize the contexts of the 11 surveys discussed in this chapter as predisposing respondents somewhat negatively toward BST, we suspect the degree of negativity is considerably milder than that to which the general public will be exposed once BST is approved for commercial use.

Results

Of course, not all of the 11 studies of BST sought the same information from their respective survey participants. Even among those who seemed to be seeking information pertaining to the same matter, the questions were often worded differently or posed in somewhat different contexts. Tables 11.2 through 11.7 represent an effort to combine questions on BST that are similar in wording or effect into manageable, meaningful categories. Some license in interpretation was necessarily exercised, but the primary purpose of this chapter is more to convey composites of the overall findings of the studies, rather than to faithfully report the specifics of any particular study. For the latter purposes, reports of the individual studies themselves should be consulted.

Consumer Awareness of BST

Except for the *DAIRY TODAY* study, all studies contained a question that somehow attempted to measure consumer awareness of BST. Table 11.2 includes the questions that were used to gauge awareness, and a summary of the responses to such questions.

The questions that related most directly to respondent's awareness of BST had different lead-ins in most cases (Table 11.1), and were each worded a bit differently (as reported in Table 11.2). In the materials available to us, the percentage of *YES* answers was sometimes the only value reported so we had to assume the balance of the answers were *NO*, although they might have actually been missing data or "don't know" responses.

However phrased, the questions pertaining to awareness of BST presumably had the same general intent--to gauge current levels of respondents' familiarity with BST. With that as given, the level of awareness among subjects in the ten studies ranged from 12.0 percent for the AHI study to 89.0 percent for the WI study. The average level of awareness was 38.0 percent.

The mean level of awareness among the three earliest of the ten studies (prior to the spring of 1990) was 23.9 percent, and the mean of the seven later studies was 44.1 percent. It is tempting to suggest that awareness has increased with time, but the ranges of awareness among the individual studies within each time group, particularly given the high awareness reported in Wisconsin, is simply too great to permit such a conclusion.

TABLE 11.2. Consumer Awareness of BST.

Study	*Question Used to Gauge Awareness*	*Participants' Responses*
AHI	Presumably "How much they had heard (or knew)" about BST.	A great deal--3.0% Some--9.0% Very little--22.0% Nothing--66.0%
PSU STATE	"Have you, by chance, already heard or read something about this?"	Yes--43.1% No--56.9%
VA	"Prior to this letter, have you read anything or heard anything about BST?"	Yes--16.6% No--83.4%
WA	"Have you ever seen, read, or heard of something called BST or bovine somatotropin?"	Yes--25.0% No--71.0% Don't Know--4.0%
NY	"Have you read anything or heard anything about BST?"	Yes--26.7% No (assumed)--73.3%
WI	"Have you heard anything about a proposed state law that would require labeling of dairy products that used milk from cows treated with BGH or BST versus those that were not?"	Yes--89.0% No (assumed)--11.0%
MO	"Have you seen or heard anything about the use of BST, bovine somatotropin, a growth hormone, in the dairy industry?"	Yes--42.0% No (assumed)--58.0%
NDB	"Which of the following issues have you recently seen or heard about in newspapers, magazines, on television or the radio?... The use of hormones to increase production of milk in dairy cows..."	Have--62.0% Have not--38.0%
PSU SCHOOL	"Have you, by chance, already heard or read something about this?"	Yes--41.9% No--57.6% No response--0.5%
PSU NATION	"Have you heard or read anything about the bovine growth hormone--it's often referred to as BST or BGH?"	Yes--22.1% No (assumed)--77.9%

Consumer Images of Milk from BST-treated Cows

Whether consumers were already aware of BST, or became aware only by virtue of their participation in one of the surveys, their feelings toward the use of BST were probably conditioned, at least in part, upon their beliefs about how the milk might be affected. Six of the 11 studies contained a question that generally related to how the consumer's image of milk might change if he/she knew it came from cows treated with BST. These questions with responses are reported in Table 11.3.

Of course, that people might think the milk will be different does not necessarily imply they think it unsafe. Nevertheless, it stretches credulity a bit to suppose that those who said they believe the milk will be different, or not the same, had in mind the milk would somehow be better than milk from cows not being administered BST. Thus, a YES response to the questions posed in the PSU-STATE and the PSU-SCHOOL studies, disagreement with the statements made in the Cornell University and National Dairy Board studies, agreement with the statement made in the Virginia study, and a *YES* to the question used in the *DAIRY TODAY* study are taken to be expressions of the general belief that milk from BST-treated cows will be inferior, or less desirable, than milk from cows not so treated.

Only in the Virginia study were there fewer respondents who looked unfavorably upon milk from BST-treated cows than were presumed to look at it favorably, or at least indifferently. Specifically, only 20.7 percent of the respondents agreed or tended to agree with the statement "The approval of BST will make milk unsafe to drink," while 35.6 percent tended to disagree or disagreed with the same statement.

If the "don't know," "unsure," "neither agrees nor disagrees," and the "no response" answers are factored out, those that remain may be presumed to imply either that BST will damage the image of milk, or that milk's image will remain unaffected. Thus, overall responses of 60.2 percent of the participants in the six studies included in Table 11.3 implied (or seemed to imply) that the image of milk will be damaged because of BST. The remainder, 39.8 percent, presumably did not think BST would affect the image of milk one way or the other. We hasten to add, however, that although the word "image" was specifically used only in the *DAIRY TODAY* study, we are employing it in a generic way in our discussion to the results of all six studies.

TABLE 11.3. Consumer Images of Milk from BST-treated Cows.

Study	*Question Used to Gauge Image*	*Participants' Responses*
PSU STATE	"Do you think milk from hormone-treated cows will be different, in some important way, from the milk being sold now?"	Yes--49.0% No--20.5% Don't know--30.5%
DAIRY TODAY	"As a customer, would the use of this new product change your image of milk?"	Yes--37.4% No--25.8% Unsure--36.8%
VA	"The approval of BST will make milk *unsafe* to drink."	Agree--7.2% Tend to agree--13.5% Tend to disagree--21.0% Disagree--14.6% Don't know--43.7%
NY	"Milk will be *safe* to drink if FDA approves BST."	Strongly disagree--7.6% Disagree--19.4% Agree--28.9% Strongly agree--3.5% Don't know--40.6%
NDB	"Milk from cows treated with BST is just like milk from untreated cows."	Disagree strongly--35.3% Disagree somewhat--27.3% Agree somewhat--17.2% Agree strongly--6.1% Neither agree nor disagree--14.1%
PSU SCHOOL	"Do you think milk from such cows will be different, in some important way, from milk that is being produced now?"	Yes--50.6% No--25.9% No response--23.5%

Consumer Feelings Toward the Use of BST

The extent to which consumers were concerned about the use of BST was synthesized from their responses to a variety of questions even more amorphous than those used to gauge consumer awareness. These are brought together in Table 11.4, nevertheless, in much the same manner that Tables 11.2 and 11.3 were developed. Our goal is to impart a general sense of the favor or disfavor expressed by consumers regarding the approval or use of BST.

Only in the Virginia study were there more people with favorable feelings toward the approval of BST than with unfavorable feelings toward its approval. The Cornell study closely paralleled the Virginia study with respect to the approach it took to this question, but more participants disagreed than agreed with the statement that BST should be approved.

The percentages of respondents who indicated they were either opposed to BST, thought it should not be approved, or were concerned about its use were averaged across all nine studies that contained such information (Table 11.4). Thus, accordingly, 56.7 percent of the respondents were found to be negatively inclined toward BST. People who professed to be positively inclined toward BST averaged 21.5 percent of all survey respondents, and the balance, 21.8 percent of the study participants, indicated they didn't know, were undecided or indifferent, or were not sure how they felt about BST.

Labeling Milk from BST-treated Cows

A question about survey participants' wishes with respect to separate labeling of milk from BST-treated herds appeared in five of the eleven studies that are reported in this chapter. There is a very strong and highly uniform level of desire for labeling across the five studies (Table 11.5). Inasmuch as milk from BST-treated cows is reputably reported to be no different than milk from nontreated cows, such labeling has become a controversial issue.

Potential Effects on Milk Consumption

Eight of the 11 reported studies focusing on the consumer side of the BST issue contained questions that provide some insight into how purchases of milk might be affected if consumers know the milk was produced by BST-treated cows. Specific quantification of the actual amounts by which consumption would change was attempted only in the studies conducted in Virginia and New York. Therefore, the information in Table 11.6 relates only to the probable direction of the

TABLE 11.4. Consumer Feelings Toward the Use of BST

Study	*Question Used to Gauge Feelings*	*Participants' Responses*
AHI	Presumably, those who said they had heard of BST were asked whether or not they favored it.	Opposed--one third In favor--one-third Didn't know--one third
DAIRY TODAY	"How concerned would you be about the use of this new product in milk?"	Very concerned--37.6% Somewhat concerned--42.9% Not concerned at all--10.6% Undecided--8.8%
VA	"BST should be approved."	Disagree/Tend to disagree--21.9% Agree/Tend to agree--44.1% Don't know--33.7%
WA	"... how would you feel about having this synthetic BST used to produce the milk you might consume? Would you be generally..."	Strongly opposed--39.0% Somewhat opposed--30.0% Indifferent--16.0% Somewhat in favor--8.0% Strongly in favor--2.0% Don't know--6.0%
NY	"BST use on dairy cows should be approved."	Disagree/Tend to disagree--32.7% Agree/Tend to agree--29.6% Don't know--37.4%
WI	"Recently state legislation has been proposed to place a temporary ban on the sale of BST in Wisconsin. How about you, do you..."	Strongly agree--40.0% Agree somewhat--24.0% Disagree somewhat--13.0% Strongly disagree--5.0% Don't know--18.0%
MO	"Would you be concerned if you learned this process involved injecting dairy cows with synthetic somatotropin which also is naturally occurring and stimulates milk production in cows?"	Yes, definitely--28.7% Yes, probably--34.3% No, probably not--17.1% No, definitely not--5.1% Not sure--14.8%
NDB	"How concerned are you personally about the use of hormones to increase production of milk in dairy cows? Are you..."	Extremely concerned--29.0% Very concerned--22.0% Somewhat concerned--39.0% Not concerned--10.0%

(continues)

TABLE 11.4. *(concluded)*

Study	*Question Used to Gauge Feelings*	*Participants' Responses*
PSU NATION	"How concerned are you about the bovine growth hormone, or BST?"	Very concerned--16.2% Somewhat concerned--40.0% Not concerned at all--15.8% Not very concerned or not sure--28.0%

TABLE 11.5. Consumer Interest in Labeling Milk from BST-treated Cows.

Study	*Question Used to Gauge Interest*	*Participants' Responses*
DAIRY TODAY	"Should milk produced with this new substance be specially labeled?"	Yes--81.0% No--7.1% No opinion and other--11.9%
VA	Milk from BST-treated cows should be labeled.	Agree/Tend to agree--85.8% Disagree/Tend to disagree--6.2% Don't know--8.0%
NY	Milk from BST-treated cows should be labeled.	Strongly agree/Agree--85.1% Strongly disagree/disagree--6.7% Don't know--8.3%
WI	"Do you support or oppose legislation that would require such labeling?"	Support--68.0% Oppose--9.0% Don't know--23.0%
PSU SCHOOL	"If this hormone is officially cleared for use by dairy farmers, do you think milk from hormone-treated herds should be labeled in such a way that it could be distinguished from milk produced by herds not using the hormone?"	Yes--82.3% No--13.0% No response--4.7%

change in consumption related by the respondents of the various surveys. The magnitude of change is uncertain, except in those cases where respondents indicated their consumption would be unaffected or that they would stop drinking milk altogether.

The question regarding consumption in the PSU-SCHOOL study related to the public school food-service directors' reluctance to serve milk from BST-treated cows to their students. Nevertheless, this study is included in Table 11.6 as though the directors were reflecting their own predisposition to drink or not to drink BST milk.

Virtually no one indicated they thought they would increase their own milk consumption because of BST. The most frequent answer was that they (the survey respondents) were unlikely to change consumption much, if at all. The average for the six of the eight studies that reported such a category was 64.3 percent.

In five of the surveys respondents were given the opportunity to indicate, if they chose, that they would stop drinking milk altogether if it came from BST-treated cows. An average of 11.3 percent of the participants in those five studies responded in that manner.

Many participants asserted their consumption would decrease in one of several different descriptive categories of change. These ranged from checking or indicating simply that they would "buy less," or their purchases would "go down" or "decrease" to "decrease a good bit" or "decrease substantially." The average of all such statements of intent to decrease their consumption of milk in such categories is 32.1 percent among the eight studies included in Table 11.6.

Because of the disparate numbers of studies that contained similar questions with summable answers as to how respondents thought BST would affect their consumption of milk, the percentages for the three categories reported above (will not change, will decrease but not stop, will stop altogether) did not add up to exactly 100 percent. Thus, the numbers reported in the text were scaled so that their sum does equal 100, and the results are shown in Table 11.7. This scaling procedure ignores the few who said their consumption would increase, and distributes the "don't know," "not sure," and "nonresponses" proportionately across the three primary categories included in the table.

TABLE 11.6. Expressed Probable Effect of BST on Milk Consumption.

Study	*Question Used to Gauge Effect*	*Participants' Responses*
PSU STATE	"How do you think knowing the milk came from hormone-treated cows would affect the amount of milk you drink?"	Increase somewhat--0.2% Not change much at all--54.7% Decrease somewhat--16.8% Decrease a good bit--11.1% Stop altogether--4.2% Don't know--13.0%
DAIRY TODAY	"How would the use of this product influence your buying of milk and other dairy products?"	Buy more--0.6% Buy less--38.0% Wouldn't change--61.4%
VA	"If BST is introduced and the price of milk stays the same, how would your weekly purchases of milk change?"	Increase slightly--1.5% Remain the same--82.7% Decrease slightly--2.3% Decrease substantially--3.1% I would stop buying milk--9.5% Don't know--1.0%
WA	"How might your future purchase of milk be affected if you knew that the milk you were buying had been produced using this synthetic BST? Would you expect that your purchases of milk would..."	Stay the same--47.0% Go down--46.0% Don't know--8.0%
NY	"If the government approved BST and the price of milk remained the same, how would your weekly purchases of milk change? Milk purchases would..."	Remain the same--82.2% Decrease--10.8% Stop--7.0%
MO	"Would you purchase milk produced by injecting dairy cows with somatotropin, a lactation stimulant?"	Yes, definitely--2.0% Yes, probably--26.0% No, probably not--24.0% No, definitely not--20.0% Not sure--28.0%
NDB	"If the milk available in stores where you shop came from cows treated with BST would you (with respect to the milk you buy for your household)..."	Increase the amount--0.0% Make no change--58.0% Decrease the amount--25.0% Stop buying it altogether--16.0% Don't know--1.0%
PSU SCHOOL	"How reluctant would you be to serve milk from such cows to your students?"	Not at all reluctant--12.4% Somewhat reluctant--44.5% Very reluctant--35.2% No response--7.9%

TABLE 11.7. Summary of Impact of BST on the Consumption of Milk

Expressed Effect on Consumption		*Average of All Respondents in the Eight Studies of Relevance (%)*
Will not change		59.7
Will decrease but not stop		29.8
Will stop altogether		10.5
	Total	100.0

Conclusions

The patterns of consumer responses over the eleven studies are remarkably consistent. The consensus that emerges from the surveys reviewed is that most consumers are concerned about BST. Most want milk from BST-supplemented cows to be so labeled, a significant number indicated they would decrease their consumption of milk if the milk came from BST-supplemented cows, and a majority said their image of milk would change if BST was involved. It is evident from these early studies that consumer reaction to BST, if introduced now, would be negative.

This negative reaction to BST is not surprising when one considers what is known about how individuals are likely to react to matters they know little about or don't understand. For example, Heiner's (1983) theory of predictable behavior suggests that when a person does not understand an issue fully, he or she is likely to fall back on rules of thumb. Generally these rules of thumb are created to avoid risk by repeating actions for which the outcome is predictable. Therefore persons are likely to resist change unless they understand a new technology (like BST) and are convinced that there is little risk involved. If not, they will continue earlier habits and routines, such as drinking milk which does not come from BST-supplemented cows.

A second possible explanation for negative consumer reaction to BST can be found in the studies of negativity bias. A central finding is that people are more likely to focus on negative characteristics and negative information when they don't know much about an issue and/or are concerned about risk (Tversky and Kahneman 1981). Studies of health and nutrition have found that consumers focus on negative information and concentrate on the negative consequences of food and health decisions (Russo et al. 1986, Hackleman 1981). It is very likely that in the early stages of the public debate concerning BST, people will be more focused and more responsive to negative than to positive information about BST.

Studies concerning nonattitudes (Converse 1970) or pseudo-opinions (Bishop et al. 1980) are also informative. The review of BST studies revealed that a majority of those interviewed were not aware of BST. The survey by Warland, et al. (1991) found that even if persons said they were aware of BST, few knew much about it. It is evident that the attitudes and perceptions of the public in regard to BST are neither well-formed nor well-founded. Studies by survey researchers indicate when a person knows very little about something, they base their attitude on general underlying dispositions (Schuman and Presser 1981). These underlying feelings or dispositions are likely to be based on positive feelings about "real" (i.e., non-BST) milk, so it can be expected that reaction to BST would be negative.

The negative reaction to BST is predictable, understandable, and real. Unless consumers are given information about BST and become convinced BST is not a threat to their well-being, these negative reactions will persist.

Acknowledgments

The assistance of the several persons who initially provided copies of the study materials and reports used as the basis for this chapter is gratefully acknowledged. Four of these individuals also reviewed an early draft of the chapter and provided useful comments and suggestions for change. These "outside" reviewers were: Anya M. McGuirk, Virginia Tech; Harry M. Kaiser, Cornell University; Robin A. Douthitt, University of Wisconsin, and Barbara Jo Slusher, University of Missouri. Their unusually high order of assistance is especially appreciated.

References

Animal Health Institute. (date unknown) "Bovine somatotropin (BST)." 119 Oronoco Street, Box 1417-D50. Alexandria, VA 22313.

Bishop, G. E., R. W. Oldendick, A. J. Tuckfarber, and S. E. Bennett. 1980. "Pseudo-opinions on public affairs." *Public Opinion Quarterly* 44:198-209.

Converse, P. E. 1970. "Attitudes and non-attitudes: Continuation of a dialogue." In E. R. Tufte (ed). *The Quantitative Analysis of Social Problems*. Reading, MA: Addison-Wesley.

Douthitt, Robin A. 1990. "Wisconsin consumer's attitudes towards bovine somatotropin (BST) and dairy product labelling." University of Wisconsin, Department of Consumer Science. Madison, WI 53706.

Fine, Travis, and Associates, Inc. 1986. *Consumer Response to the Introduction of BST Technology: Devising a Communications Strategy. Final Report.* Report prepared for the National Dairy Board, 2111 Wilson Boulevard, Arlington, VA 22201.

Hackleman, E. 1981. "Food label information: What consumers say they want and what they need." *Advances in Consumer Research* 8:477-478.

Heiner, Ronald A. 1983. "The origins of predictable behavior." *American Economic Review* 74:560-595.

Henderson, Pam C. 1989. "Consumers balk at BST." Philadelphia, PA: *Dairy Today*. November/December.

Kaiser, H. M., C. W. Scherer, and C. M. Barbano. 1991. "Consumer perceptions and attitudes towards bovine somatotropin." Paper submitted for publication in *Northeastern Journal of Agricultural and Resource Economics*.

McGuirk, Anya M., Warren P. Preston, and Gerald M. Jones. 1990. "Biotechnology and the consumer: the case of BST." SP-90-6. Blacksburg, VA: Virginia Polytechnic Institute and State University, Department of Agricultural Economics.

Norris and Associates. 1990. "Washington Dairy Products Commission BST statewide consumer survey Wave II." 17111 161-st Avenue, N.E., Woodinville, WA 98072.

Preston, Warren P., Anya M. McGuirk, and Gerald M. Jones. 1991. "Consumer reaction to the introduction of bovine somatotropin." In Julie A. Caswell (ed). *The Economics of Food Safety*. New York, NY: Elsevier Science Publishing Company Inc. Chapter 10.

Russo, J., R. Staelin, C. Nolan, G. Russell, and B. Metcalf. 1986. "Nutrition information in the supermarket." *Journal of Consumer Research* 13:48-70.

Schuman, H., and S. Presser. 1981. *Questions and Answers in Attitude Surveys*. New York, NY: Academic Press.

Slusher, Barbara J. 1991. "Consumer acceptance of food production innovations--an empirical focus on biotechnology and BST." In Robert N. Mayer (ed). *Enhancing Consumer Choice*. Columbia, MO: American Council on Consumer Interests.

Smith, Blair J. 1991. "Consumer and public school perspectives on the BST issue in Pennsylvania." *Farm Economics*. March/April. University Park, PA: Pennsylvania State University, Cooperative Extension Service.

The Research Alliance. 1990. "Consumer reactions to the use of BST in dairy cows." San Francisco, CA: Marketing Research and Counseling Services.

Tversky, A., and D. Kahneman. 1981. "The framing of decisions and psychology of choice." *Science* 211:453-458.

Warland, Sterngold, and Herrmann. 1991 (forthcoming). "American Knowledge and Concerns About BST." University Park, PA: Department of Agricultural Economics and Rural Sociology.

12

Food Safety and Product Quality

Manfred Kroger

It is quite understandable that there is public discussion in regard to costs and benefits of using BST. After all, we live in a democratic society, a society that is reasonably literate and that has invented for itself *consumerism*. This is understood to be a mechanism or process, uncontrolled and haphazard as it may be, that rightfully examines the interactions between and among industry, governments (the regulators and consumer protectors) and the various publics.

The phenomenon of the questioning public, via consumer advocates, is relatively new. Ralph Nader's emergence in the 1960s to become the eminent spokesperson for the American public needs to be recognized. Meanwhile the debate to determine if consumer activists are obstacles in the path of progress and inventiveness or do they truly and unselfishly try to protect the public (and the environment, and the Earth as a whole) continues. A facetious critic might say:

> Where were these protectors of the common good when gunpowder, or weapons, or the automobile were invented, as these are things that maim and destroy and can not possibly be of any good if one extrapolates their effect on human health well into the future.

The problem with inventions nowadays, or with anything new in technology, is not only to satisfy the hurdles of gaining legislative and regulatory approval for use--it includes convincing consumers and consumerists alike of the societal benefits and possible risks that will follow in the wake of adoption of the new idea.

Risk communication is extremely difficult. This is known by parents as well as by communicators of modern-day technology. A child's fear of the dark can literally be dissolved by the process of illumination. However, the perceived fear of an intellectual is another matter. Heaps of data on documented benefits and the practical absence of risk are not enough to undo an old belief.

BST treatment, a new technology capable of getting 7 to 23% more milk from cows, is now a reality. The question of what daily BST injections will do to a cow's long-term well-being is not addressed in this chapter; neither is the economic uncertainty of what it might do to the milk-surplus problem or the dynamics of American rural life. What is addressed here are the questions: "Is milk from BST-treated cows safe to consume?" and "When cows are made into hamburger, can that meat be eaten without ill effects on human health?"

The question of safety (as well as the question of efficacy) must be addressed by the Food and Drug Administration (FDA). The safety of foods from cows in BST research was determined by FDA in 1985. Summaries of studies supporting the safety of milk and meat from cows receiving BST have been published by FDA researchers Juskevich and Guyer (1990) under the title: "Bovine Growth Hormone: Human Food Safety Evaluation" and by Daughaday and Barbano (1990) under the title: "Bovine Somatotropin Supplementation of Dairy Cows--Is the Milk Safe?" It is very appropriate to reprint the abstracts of these two reports here. The FDA authors state that:

> Scientists in the Food and Drug Administration (FDA), after reviewing the scientific literature and evaluating studies conducted by pharmaceutical companies, have concluded that the use of recombinant bovine growth hormone (rbGH) in dairy cattle presents no increased health risk to consumers. Bovine GH is not biologically active in humans, and oral toxicity studies have demonstrated that rbGH is not orally active in rats, a species responsive to parenterally administered bGH. Recombinant bGH treatment produces an increase in the concentration of insulin-like growth factor-I (IGF-I) in cow's milk. However, oral toxicity studies have shown that bovine IGF-I lacks oral activity in rats. Additionally, the concentration of IGF-I in milk of rbGH-treated cows is within the normal physiological range found in human breast milk, and IGF-I is denatured under conditions used to process cow's milk for infant formula. On the basis of estimates of the amount of protein absorbed intact in humans and the concentration of IGF-I in cow's milk during rbGH treatment, biologically significant levels of intact IGF-I would not be absorbed (Juskevich and Guyer 1990, p 875).

Daughaday (Washington University School of Medicine, St. Louis, MO) and Barbano (Department of Food Science, Cornell University, Ithaca, NY) summarized their paper as follows:

> Complex, biologically active proteins (eg, enzymes and hormones) can be manufactured safely and cost-effectively through applications of biotechnology. Some of these proteins (eg, human insulin, human somatotropin, rennet for cheese manufacture) are currently approved for medical or food processing applications. Bovine somatotropin (BST) for lactating dairy cattle is another product that can be produced via biotechnology and may allow dairy farmers to produce milk at a lower cost. In 1985, based on an evaluation of toxicological data, the Food and Drug Administration concluded that milk and meat from BST-supplemented cows was safe and wholesome. The Food and Drug Administration has authorized the use of milk and meat from BST-supplemented cows in the commercial food supply. Its evaluation of the impact of BST supplementation on the long-term health of dairy cattle is near completion, and BST may be approved for commercial use in early 1991 (Daughaday and Barbano 1990, p 1003).

The conclusions of this latter report also are worth quoting (p 1005):

1. The FDA has answered all questions and concerns about the safety of milk from BST-supplemented cows for human consumption and has authorized the commingling of milk from the BST-supplemented investigational herds with the rest of the commercial milk supply.
2. Bovine somatotropin has no biological activity in humans when ingested orally or when given by intramuscular injection.
3. Insulin-like growth factor I is not orally active. Any changes in IGF-I levels in milk are well within normal variation and are lower than those reported in human milk.
4. All cow's milk contains BST, and no significant change in milk composition occurs as a result of giving cows supplemental BST.

These journals are highly respected, rigorously edited, peer-reviewed publications. The evidence presented in the two articles is based on several dozen relevant primary research papers. The objective here is not to recite and re-review, but merely to report what is known about the safety of BST based on reliable research.

In addition, a National Institutes of Health panel of eleven medical and veterinary experts, after assessing the scientific literature on BST, concluded that milk and meat of cows treated with the hormone are as safe as those from untreated cows. Specifically, the composition of milk produced either way is essentially identical, the only difference being an increase in yield (output per cow) by roughly 10 percent. However, the panel recommended further research on the human-health effects of one component, found at slightly higher concentrations in the milk of

BST-treated cows when compared with that of nontreated cows, namely, IGF-I (insulin-like growth factor-I).

So far, there are only two groups that seem to question or oppose the use of BST: the Foundation for Economic Trends, a Washington-based consumer/environment protection interest group, and small dairy farmers. The former has a record of opposition to genetic engineering in general; the latter fear the new technique will benefit only large milking operations.

Both objections are based on conjectured negative outcomes and do not address the facts of *absence of risk* to public health. Neither group seems to acknowledge that, overall, societal good will most likely be generated by BST technology.

The Foundation for Economic Trends deserves credit for asking the question as to harm that may come to the dairy cow as a result of BST administrations. It seems that the animal rights/animal welfare issue ought to be dealt with in another forum. But BST use can well be anticipated as inconsequential to this issue, or at least no more severe than other conventional animal-management procedures.

The *New York Times*, known to be a liberal newspaper speaking for the rights of the oppressed and exploited, came out *for* BST, chiding opponents of BST in an editorial of May 1, 1990. It made these points worth noting:

1. The synthetic hormone injected into the cow is almost identical to the cow's natural hormone; small amounts of which naturally occur in the milk.
2. Even if the growth-hormone level in milk of treated cows is slightly elevated, the compound is destroyed just like any other protein passing through the human digestive system.
3. Natural or synthetic bovine growth hormone differs so greatly from human growth hormone that it would have no biological effect, should it somehow enter the human bloodstream.

It is of interest to nutritionists, dieticians, cheesemakers, and the milk-processing industry in general to know about compositional changes in milk obtained from BST-treated cows. The issue of changes in meat composition can be considered subordinate to the milk issue. In other words, if there are any noteworthy product alterations as a result of daily BST treatment, the effect would first be evident in the milk, rather than in the meat at the end of a cow's life and after a period of presumed nontreatment prior to slaughter.

Research has been conducted on various specific topics during the 1980s. The results have not been unusual and do not merit restatement here, except in outline form. Milk composition in general is far more subject to diet, state of health of cow, inherent genetic factors, and environmental conditions than it is to BST treatment.

Studies have been conducted on BST administration and its effect on various components:

- There are slight increases and decreases, respectively, in milkfat and milk-protein content immediately after treatment, as would be expected, and as is common after any feed or metabolic adjustment.
- Milkfat, protein, lactose, total solids, and solids-not-fat percentages are, overall, unaffected over a full lactation period and not different from those of milk from nontreated cows.
- Milk-ash or mineral content, specifically phosphorus and calcium contents, are not altered by BST treatment.
- A slight shift in the Kjeldahl fractions (casein, whey protein, and nonprotein nitrogen) have been observed in some experiments. This is of no consequence to milk quality, but may benefit cheesemakers if cheese yields are improved.
- There are no effects on the relative proportions of short-, medium-, and long-chain fatty acids. If milkfat could be made more unsaturated by BST treatment, it would be a major breakthrough of immense nutritional significance.
- Changes in free fatty-acid content have not been noted; therefore influence on off-flavor "rancidity" is not anticipated, nor is vulnerability to oxidized flavor development.
- Other studies in progress will most likely show inconsequential results. Examples include effect on milk-cholesterol content, physical characteristics of milkfat, bacterial growth, antibiotic-testing methods, and cheese manufacture.

Regretably, the information presented in this chapter has omitted considerable detail, but such is the nature of risk communication. I urge all interested in BST and its effects on the dairy industry--critic and proponent alike--to obtain and study the research publications mentioned herein, as well as the papers upon which they are based.

The scientists cited are committed to follow future developments on BST and will continue to be most appropriate as resource persons.

References

Juskevich, J. C., and C. G. Guyer. 1990. "Bovine growth hormone: human food safety evaluation." *Science* 244:875-884.

Daughaday, W. H., and D. M. Barbano. 1990. "Bovine somatotropin supplementation of dairy cows--is the milk safe?" *Journal of the American Medical Association* 264(8):1003-1005.

13

BST and International Agricultural Trade and Policy

Harald von Witzke and Claus-Hennig Hanf

A prime concern of mankind, throughout history, has been to secure the supply of a sufficient quantity of food. Beyond the main components of food such as calories, protein, or fiber, quality considerations played only a secondary role. For a long time, food was considered by economists as ultimately constraining economic development and welfare (Malthus 1798, Ricardo 1895). In fact, it was food pessimism that earned economics the name "dismal science."

This pessimistic view of growth and development was very popular until well into the 1970s. By the early 1980s, however, it was replaced by concerns over food surpluses in the developed world and their economic and political consequences (von Witzke and Ruttan 1989).

The realization that the earth's potential to produce a much higher quantity of food than is demanded now or will be demanded in the future (provided current projections of global population growth are correct) is paralleled by a growing emphasis on food-quality components.

The use of growth hormones in the production of dairy and beef is considered by many consumers to have a negative effect on quality. The implications for international agricultural trade and policy resulting from the growing demand for food quality can be nicely demonstrated by lessons learned from the use of growth hormones which was one of the first food-quality components to cause a major international trade dispute.

It can be expected that the growing demand for food quality will increasingly affect international agricultural trade relations as well as international trade policy. An analysis of the economic implications of BST regulation on these and related issues can help to improve our understanding of the international economic and political implications resulting from the growing demand for food quality in general.

In this chapter we will first analyze the problem of food quality and BST in terms of market failure. Then, we will discuss the special case of BST. Third, we will analyze the impact of BST regulation on social welfare. Fourth, we will derive implications of BST regulation for international agricultural trade and policy in light of our theoretical analysis.

Food Quality and the Market

Dimensions of Food Quality

Any good possesses an infinitely large number of properties (such as size, weight, smell, color, durability, freshness, and date produced or presented) that may influence its (perceived) quality. Here we define quality as the aggregate of all properties that affect the decisions of economic agents.

Following Nelson (1970), we distinguish between (i) search goods, and (ii) experience goods.

"Search goods" have properties that are measurable and recognizable by consumers before purchase; the properties of "experience goods" can be recognized only after purchase.

For the purpose of analysis, we added a third category, namely (iii) goods with indeterminable quality components (Hanf and Wright 1990). The quality components of these goods cannot be determined by the consumer. The relevant quality components of this category of goods are usually determined by the mode of production. An example would be goods produced following certain rules of animal rights organizations. As argued later, this category includes dairy and beef produced using growth hormones.

Food Quality and Asymmetric Information

Economists typically think of a market in which quantities of a well-defined good are exchanged, for a price, between buyers and sellers. Much in the same way one can think of a market for the quality components of a given good. On a perfectly competitive market, sellers would supply alternative qualities of a good and consumers would demand them. The laws of supply and demand would generate equilibrium prices for various qualities of a good. The higher the (perceived) quality of the good, the higher the price.

In the real world, however, markets for quality components are not perfectly competitive. As Ackerloff (1970) emphasizes, on many markets for quality, there is asymmetric information between buyers and sellers. Typically in these cases buyers have less information than sellers (see also Klein and Leffler 1981, Kinsey 1990, Kramer 1990). Asymmetric information affects markets of experience goods and of goods with indeterminable quality components, but for obvious reasons asymmetric information does not play a role on markets of search goods.

Food Quality and Market Failure

The market mechanism has a tendency for self-correction of failure due to initial asymmetric information between buyers and sellers. Therefore, the consensus among economists is that many markets for quality components do not require government regulation to optimally allocate scarce goods and resources. However, the market mechanism fails to function properly with regard to goods with indeterminable quality components, including goods which may impose health risks on consumers. In these cases, government regulation is required for optimal allocation of factors and goods.

The economic reasoning behind this possible market failure can be illustrated by the following example. Assume a dairy good that is sold by producers directly to consumers. There are many producers and many consumers, thus no single producer or consumer can actively influence the price that forms on the market through supply and demand. Initially, there is no government regulation of this market.

The dairy good comes in two distinct qualities but it is otherwise homogenous. Quality A is produced using a feed additive (residues of which show up in the good), and consumers believe that this results in a health risk later in the lives of those who consume the good. Quality B is produced without the use of the feed additive. The marginal costs of producing quality A are lower than those of producing quality B. Market forces generate an equilibrium price for each quality, such that $P_A < P_B$, where the difference between both prices represents the differences in marginal costs and the consumers' willingness to pay for the quality component, i.e., their willingness to pay for avoiding the health risk resulting from the consumption of quality A.

For the honest individual producer, it is then straightforward to determine the quantities that maximize profits on the basis of market prices for those two qualities of the good and the cost of producing each quality. The producer, of course, has information on which quantities of the good belong to each of the two quality categories, but consumers do not unless they have each unit of the good tested for residues of a chemical agent. Of course, in real life, this would be very expensive or otherwise not practical. As the marginal cost of producing a unit of quality A are lower and the price of a unit of quality B is higher, and as consumers cannot tell the difference between both qualities, there is an incentive for producers to cheat by producing quality A and selling it as the higher-priced quality B.

Consumers are not assured that they actually receive quality B in exchange for paying P_B. In fact, it is reasonable for consumers to expect that there is a high probability that they will be sold quality A, even when paying the higher price. Therefore, consumers will demand only quality A. Quality B will not be produced. Although consumers may be willing to pay the higher price for quality B, the market mechanism fails to actually supply this good. The central reason for this is that consumers have no practical way of finding out immediately whether the good incorporates the quality component paid for or not.

Self-correcting Market Failure and Government Regulation

Notice, however, that markets tend to function when consumers have the possibility of acquiring information on a quality component purchased at reasonable cost even if they cannot judge its quality at the time of purchase. This could be the case when a sensory quality is promised which cannot be verified at the time of purchase. Should consumers find out later that an advertized quality component is nonexistent they will not buy the good again from this producer. On the other hand, if a producer consistently offers the quality component he or she builds a reputation as a reliable supplier of the quality component and thus provides consumers with the assurance that they actually receive the quality they pay for. In this case, the market mechanism self-corrects, via a repeat-purchase process, its initial failure at generating information on quality. Obviously, this self-correcting mechanism fails in the case of BST.

When markets for quality fail because of asymmetric information, government regulation can improve the functioning of markets. There are several basic instruments that can be employed, such as labeling requirements or maximum standards for residues. Here we refer to such instruments as food-quality standards.

Government regulation of food quality must include the definition of labeling requirements for food-quality standards, monitoring of compliance, and enforcement of compliance in order to assure consumers that they actually receive the quality they pay for. Of course, the administrative costs of government regulation of food quality can be quite high, particularly when a large number of producers have to be monitored and when the food chain is long. Incentives to offer a lower quality good at the price of the higher quality good exist at every level of the food chain.

BST Market Performance and Government Regulation

The public and political debate on BST for commercial use in dairy production suggests that BST use will become subject to some form of government regulation. Government regulation of BST is endogenous and a function of a number of variables, such as level of economic development, the political economic as well as the microeconomic environment, or a country's dependence on international trade. Of course, the more that countries differ relative to these variables, the more are the international differences in government regulation of BST and the subsequent international trade distortions.

The Public BST Debate

The potential adoption and use of BST has stirred considerable public debate. As soon as the first news of BST experiments at agricultural research stations and universities appeared in the media, considerable public concern over its use was expressed in many countries. As food quality is a luxury good (Falconi and Roe 1991), it is not surprising that the public debate over BST was much more pronounced in developed than in less-developed countries.

But even among high-income countries there have been significant differences in the intensity of the public debate. Discussion in the United States has emphasized possible human health risks. The BST issue has been more controversial in Europe, where it included a wide variety of concerns besides human health. Animal rights groups criticized BST use as incompatible with their basic principles; and environmental groups objected to BST because of their fear that it would lead to a growing concentration of animal production.

Moreover, technology pessimism appears to be more prevalent in Europe. Many Europeans distrust scientifics' assertions of new technologies as being totally safe and without negative side effects. In part, this is due to recent developments where advances in medical science led to reversals in the assessment of many drugs and technologies such as asbestos, garbage incineration, or certain pesticides (Huisinga 1989). In Europe, biotechnological innovations based on gene technology are met by even more pronounced skepticism (Backhaus 1989); of course, this includes BST. Many Europeans doubt, moreover, that in view of persistent agricultural surplus production, further production-enhancing technologies are desirable (OECD 1988). The differences in the public perception of BST in the United States and in Europe are likely to lead to differences in the regulation of BST use, and thus are likely to affect international agricultural trade relations between political entities on both sides of the Atlantic.

Consumer and Producer Response

Producer and consumer opinions regarding BST were discussed by Fallert and Hallberg in Chapter 10, and by Smith and Warland in Chapter 11. Therefore, our treatment of these topics can be very brief.

The public debate over BST use in dairy production suggests that consumers will not consider dairy products or beef produced using growth hormones as being of superior quality. There is a significant group of consumers in high-income countries who consider such foods as being inferior. Hence the demand curves for conventional dairy and beef products will be to the right of the respective demand curves for the same goods produced with growth hormones. The marginal costs of production with BST are lower than those of conventional production. Estimates of the farm-level economic effects of BST use suggest a reduction in cost of dairy production by 5 to 10 percent (e.g., Brown 1988, Fallert et al. 1987, Marion and Wills 1990, Schmidt 1989, Zeddies and Dolutschitz 1988, Zepeda 1988). The resulting shift in supply, together with the expected demand shift, would in a closed-economy framework lead to market prices of BST-produced goods that would be lower than those of conventional goods. Whether intercountry differences in preferences and BST-related cost reductions result in intercountry price differences depends on transportation cost and dairy policies employed.

As long as some consumers prefer lower-priced BST over conventional dairy products while others are willing to pay a higher price for conventional dairy products, social welfare may be positively affected when both qualities are supplied (see below for details). However, as discussed earlier, BST goods incorporate an indeterminable quality component. Hence, markets will fail to supply both qualities and government intervention is called for. Obviously, a labeling requirement has to be introduced. In order to provide consumers with the necessary assurance it is also necessary to monitor and enforce compliance on all levels of the food chain (including production, processing, transportation, retailing, etc.).

Given the nature of BST goods, monitoring and enforcement of compliance will be very costly. The cost of monitoring and enforcement will increase the cost of supplying the conventional dairy product, resulting in price increases of conventional dairy products. Of course, if the market share of conventional dairy products shrinks after BST approval there may be additional price increases due to diseconomics of scale.

BST and Social Welfare

It is sometimes argued that the use of BST will always act to increase social welfare, as it reduces marginal costs and widens consumers' choices. Implicitly, this argument assumes (a) that BST use does not involve a human health risk; (b) that the costs of producing BST goods are generally less than those of producing conventional goods, and (c) that BST use does not require additional resource use for processing, transportation, handling, etc.

As we shall discuss below, the direction and magnitude of the social welfare effect of BST use also depends on consumer preferences and their ability to actually distinguish between a conventional and a BST good. Assume a closed economy (an economy that does not engage in dairy product trade). Let the good be fresh milk, and assume further that the production costs of conventional milk are higher than those for BST milk. Now let us distinguish between the following cases:

1. Consumers are indifferent with regard to conventional and BST milk.
2. Consumer preferences are heterogenous. Some consumers prefer conventional milk over BST milk and they are able to distinguish conventional from BST milk.
3. Consumer preferences are heterogenous. Some consumers prefer conventional milk over BST milk but are unable to distinguish conventional milk from BST milk.

In the first case, there will be a social welfare gain, as a given quantity of a good considered to be homogenous by consumers is produced at lower cost when BST is used. In the second case, the social welfare effect is unambiguously positive as well; the welfare position of consumers who prefer conventional milk does not worsen while everybody else is better off because the price of BST milk is less than that of conventional milk.

In the third case, the social welfare effect is ambiguous, however. To demonstrate this assume a linear demand function for conventional milk which prevailed prior to the introduction of BST:

$$(1)\ P_d = A - ax$$

where

P_d = price of conventional milk,
x = quantity of conventional milk, and
A, a = parameters of the demand function.

For reasons discussed above, after the introduction of BST in dairy production all consumers will behave as if all milk were BST milk, assuming no regulations are introduced or the regulations are perceived as ineffective. Of course this will act to shift the demand function to the left. Let this demand function be:

$$(2)\ P_{d1} = A - \alpha - ax_1$$

where

P_{d1} = price of BST milk,
x_1 = quantity of BST milk, and
α = shift parameter.

Now assume that the marginal cost (= supply) function for conventional milk is

$$(3)\ P_s = B + bx.$$

The introduction of BST will shift the marginal cost curve downward. Let this marginal cost curve be:

$$(4)\ P_{s1} = B - \beta + bx_1.$$

Before the introduction of BST, market clearing implies:

$$(5)\ x = (A - B)/(a + b)$$

The market equilibrium after introduction of BST is:

$$(6)\ x_1 = (A - B - \alpha + \beta)/(a + b).$$

The change in social welfare (D) resulting from the use of BST is the difference in social welfare in both situations (W and W_1):

$$(7)\ D = W_1 - W.$$

Social welfare is measured as the sum of producer and consumer surplus. Hence we obtain:

$$(8)\ W = \{0.5\ (A - B)^2\}/(a + b),$$
$$(9)\ W_1 = \{0.5\ (A - B - \alpha + \beta)^2\}/(a + b), \text{ and}$$
$$(10)\ D = \{0.5\ (a + b)\}\ \{(\alpha - \beta)^2 - 2(A - B)\ (\alpha - \beta)\}.$$

Equation (10) implies:

$$D > 0 \text{ if } \beta > \alpha,$$

$$D = 0 \text{ if } \beta = \alpha, \text{ and}$$
$$D < 0 \text{ if } \beta < \alpha < 2(A - B) + \beta.$$

This result implies that the social welfare effect is more likely to be positive (negative) the larger (smaller) the reduction in marginal cost of production caused by the use of BST and the less (more) aversely consumers react to the introduction of BST milk.

Consumer aversion to BST milk tends to grow with increasing income, while the reduction in marginal cost of milk production is higher the larger is herd size and the higher the human capital input. Therefore, high-income countries (such as those of the European Community) with relative strong consumer aversion to BST and with relatively low efficiency of production due to small herd sizes (and perhaps less skilled labor), would be worse off from use of BST. The change in social welfare of a country such as the United States with relative efficient dairy production and some consumer aversion cannot be determined theoretically. In less-developed countries where both the efficiency of dairy production and consumer incomes are low, the change in social welfare cannot be determined theoretically either. However, to the extent that consumers in these countries can be expected to be indifferent with regard to the kind of milk they can buy, and if there are at least some efficiency gains from BST use, social welfare in those less-developed countries will increase through the approval of BST.

Producer Interests

Policy decisions affecting agriculture reflect the activities and relative influence of special interest groups involved. Agricultural-producer groups tend not to be very powerful in less-developed countries, but they are quite influential in high-income countries. Hence, farmers' interests and their relative importance in the policy-making process can help shed further light on government regulation of BST.

In a closed-economy framework where international trade in dairy products is limited, the impact of BST on producer incomes may be negative or positive. In developed countries the demand for dairy products at the producer level is inelastic. Hence, producer income would decline with BST use, all other things being equal. In many low-income countries the demand is elastic; consequently, BST use would lead to higher producer incomes, all other things being equal.

In an open-economy framework, however, farmers would generally favor BST use as long as they expect other countries to approve BST, so as to remain competitive. In fact in an open-economy framework farmers may push for quick approval of BST in order to capture rents of early technology adoption.

International trade in dairy products, relative to total world production, is very limited. This suggests that a closed-economy framework may be adequate to analyze supply, demand, and market price reactions. To the extent that this is the case, one would expect that producers in less-developed countries would favor BST use while those in developed countries may oppose it. In less-developed countries, consumers have relatively more political economic market power. As the demand for food quality is low at low levels of economic development, consumers would tend to prefer the introduction of BST (as would producers). In developed countries, agricultural producer groups are more powerful than are consumers. As already mentioned, farmers' interest groups would be averse to BST use. Again there would be a coincidence of interests between producers and consumers, because with increasing income the demand for food quality increases.

Of course, agricultural and trade policies also influence farmers' attitudes towards BST approval. This can be illustrated using European Community dairy policy as an example. Both producer price and quantity are regulated by the government through a system of price support and domestic production quotas. The quota for the European Community is such that, at the support-price level, domestic production exceeds domestic consumption. The surplus production is exported, and an export subsidy is paid by the European Community. The introduction of BST would reduce production costs and thus increase producer incomes. Hence, it is not surprising that European Community farmers support approval of BST.

Prospects for BST Approval and Government Regulation

In some countries (e.g., Soviet Union, Czechoslovakia) the use of BST has already been approved, but many other countries are quite hesitant to progress with procedures that would lead to a legal use of

BST. The United States is far along in approving BST use. The Food and Drug Administration's tests are finished and final approval is expected for 1991 (Juskevich and Guyer 1990).

In the European Community, a final approval of BST has been postponed repeatedly. In the 1980s, American BST manufacturers have repeatedly demanded, of the European Community Commission, approval of BST for commercial use in dairy production (Biotechnik Spezial 1988). After several delays a final decision by the Council of Ministers was scheduled for December 1990. But again, the decision was postponed. The central reason for these delays is the dilemma facing the European Community. If the use of BST is approved, the Community can expect strong political disputes with consumers and environmental groups, both very vocal in their opposition to BST. Should BST use be prohibited, the Community would face opposition from dairy producers and in turn be confronted with international trade disputes with the United States.

We know of no country where deregulation of BST use is seriously contemplated. Most countries have in place, or are presently developing, a combination of the following regulatory instruments.

Regulation of BST Application. Most countries are considering procedures designed to avoid (or at least minimize) human health risks of BST use and to reduce consumer concerns. The main problem of this type of BST regulation is that monitoring and enforcement are very costly. If intercountry differences in BST-application procedures exist, they can be used as nontariff barriers to trade. In some cases it even may be used for a ban of all dairy products originating in a country with weaker regulations.

Maximum Residue Standards. Another regulatory instrument under discussion in many countries is maximum residue standards for BST goods. The reason for this is that very high levels of synthetic growth hormones have caused measurable health effects in laboratory animals (Juskevich and Guyer 1990). If there are intercountry differences in BST-residue standards, they can likewise act as nontariff trade barriers. However, maximum-residue standards do not address concerns regarding proper application on the farm and those related to animal welfare.

Labeling and Separation. If BST is used properly, the quality of those goods cannot be distinguished from conventional goods with known scientific methods. Since some consumers are opposed to BST use, labeling of BST goods is under consideration in many countries.

However, labeling is not likely to provide the necessary assurance of food quality unless there is a suitable monitoring and enforcement system in place.

Because of the problems of distinguishing between BST and conventional goods, procedures that would keep both types of products apart (in the production/processing/marketing chain) from production to final consumption are being contemplated as well. Again, intercountry differences in labeling and separation requirements can act as nontariff barriers to international trade.

BST and International Agricultural Trade and Policy

The demand for food quality as well as for food-quality regulation are functions of a number of variables. In this section, we will first discuss the major variables determining the demand for food-quality regulation and analyze how food-quality regulation can act as a barrier to international trade in food and agricultural commodities. Then we will discuss the political economic aspects of food-quality regulation in an international trade context.

Food-quality Regulations as Trade Barriers

Growing incomes, together with improved knowledge about health risks, have led to a significant growth in demand for food quality and, with a time lag, for food-quality regulation in the last decade. Of course, the demand for food quality and food-quality regulation are also functions of other environmental hazards that consumers are exposed to, as well as of (nationally divergent) preferences.

Moreover, the demand for food-quality regulation is driven by the growing opportunity costs of human time. The costs of obtaining information on food-quality components tend to rise with increasing opportunity costs of time. In addition, rising opportunity costs of time stimulate increased demand for food away from home (Senauer 1979), and thus lead to a growing intake of food for which there is uncertainty about quality components (Falconi and Roe 1991).

As the combination of variables which determine the demand for food quality and for food-quality standards differ from one country to another, there may be differences in each country's quality standards (in terms of positive as well as normative analysis) in the absence of international policy coordination. *De facto*, such nationally divergent standards represent barriers to trade.

Obviously, the trade effects of nationally diverging food-quality regulations depend on the mix of regulatory instruments implemented by each country and on the intercountry differences in food-quality regulation. Of the instruments discussed in this paper, labeling of BST goods would result in the least amount of trade distortions while prohibiting imports and sale of BST goods altogether would obviously be the most restrictive trade barrier.

If each country sets its own standards but will accept those of other countries as well, as long as the products are clearly labeled, there may be a range of qualities of goods that can be freely traded. This may be beneficial if consumer preferences for food and quality are heterogenous. A range of differing product standards for the same basic type of good, however, interferes with the signalling mechanism of a particular country's set of standards and serves to increase consumer uncertainty. Information has a price.

More importantly, however, food-quality standards can well be defined such that only one quality can legally be sold, be the good domestic or foreign in origin. This was the case in the beef dispute between the European Community and the United States (Kelch 1988). The European Community had banned the sale of all beef that may have been produced using growth hormones. As there is no method to determine if growth hormones were used in producing beef, and because use of growth hormones is legal in the United States, the European Community simply banned all beef imports from the United States, thus converting a domestic food-quality standard into a prohibitively high nontariff trade barrier.

Such nontariff trade barriers, by their very nature, can easily give rise to international trade policy disputes because it is very difficult to tell when such instruments are used to protect domestic consumers from health hazards and when they are intentionally used to protect domestic producers (Kinsey and Houck 1991). Certainly, nontariff trade barriers, many based on food-quality standards, are on the rise in international food and agriculture. In 1986, about 75 percent of U.S. food imports in value terms and 45 percent of raw agricultural commodity imports in value terms were subject to some form of nontariff trade barriers. For the European Community these numbers were 100 percent and almost 30 percent, respectively (Laird and Yeats 1990).

Barriers to trade in the form of food-quality standards could be removed through international harmonization of food-quality standards. An international agricultural trading system free of barriers represents an international public good. A single country cannot supply itself with such a good except in cooperation with other countries (Runge, von Witzke, and Thompson 1989, Runge 1991). To the extent that a country can be made better-off by internationally harmonized food-quality standards there is an incentive for it to pursue political strategies which could lead to harmonization of food-quality standards among countries.

However, international harmonization of food-quality standards may be very difficult to achieve for a variety of reasons. First, the general problems of public goods provisions have to be solved. As is well known, public goods are difficult to provide efficiently because of incentives for free-riding by the agents involved. Moreover, agreement has to be achieved over the distribution of the costs of providing the public good.

Second, the international distributive problems are aggravated in the case of food-quality standards because demand for both food quality and food-quality standards are affected by such variables as incomes, information, or preferences. However, these variables are rather unequally distributed among countries, making it more difficult to agree on uniform food-quality standards.

The growing importance of food-quality standards will also redefine trade relations between the developed and the developing countries. On one hand, food-exporting less-developed countries may face more barriers to trade, as developed countries tighten existing food-quality standards and introduce new ones. Less-developed countries may have problems meeting standards set by wealthy countries because, in many cases, new production techniques are human-capital intensive. Human capital, however, is scarce in most low-income countries. Moreover, production technologies that meet the standards developed in wealthy countries may not be efficient under the climatic conditions or relative factor prices in developing countries. On the other hand, food-exporting less-developed countries may also be the beneficiaries of tighter food-quality standards in high-income countries, as many of them have traditionally employed production techniques using fewer chemical inputs. Yet another dimension of this problem is that some food-exporting developed nations may opt to constrain domestic sales of foods to those products that meet domestic standards but allow the production of below-standard foods for export to other countries with less-tight regulation.

Political Economic Aspects of Food Quality, BST Regulation, and International Trade Policy

Food-quality considerations and food-quality standards will play an increasingly important role as nontariff barriers to trade for political economic reasons as well (see also Runge 1991). The growing demand for food quality tends to favor political coalitions between farm interest groups and consumers. In developed countries, the influential minority of agricultural producers, seeking protection from foreign competition, may find increasingly attractive coalition partners in consumers seeking protection from food-related health risks via food-quality standards. Political demands of both groups can be met by using domestic food-quality regulation as a nontariff barrier to trade.

Food-quality standards may gain in importance also as a substitute for traditional tariff or nontariff barriers to trade. Progress in international negotiations on a more liberal international agricultural trade regime, such as those in the GATT, may have been slow but they have contributed to significant policy adjustments in both the United States and the European Community. Agricultural price supports in real terms have been reduced remarkably in the United States and the European Community in recent years. If traditional forms of agricultural income support continue to lose their importance, agricultural interest groups will try to find substitute supports. Again, food-quality standards and a coalition with consumers is likely to be an attractive option.

The dispute between the European Community and the United States over shipments of beef that may have been produced using hormones points in this direction. In the European Community, beef production has increased. The degree of self-sufficiency in beef is above 100 percent and budgetary expenditures for beef price support have grown significantly. Budgetary expenditures appear to be an important determinant of price support in most developed countries (see e.g., Riethmuller and Roe 1986, von Witzke 1986, 1990). Hence, keeping foreign beef out of the European Community can be seen not only as an attempt by domestic producers to expand domestic sales at the expense of foreign competitors but also to maintain a higher level of support.

Summary and Conclusion

Using BST as an example, we have attempted to analyze the implications for international agricultural trade and policy of the growing demand for food quality. This demand is very likely to result in an increased proliferation of (nationally divergent) food-quality standards. Many of them will, *de facto*, act as barriers to trade. Political coalitions between consumers demanding protection, through food-quality standards, from health risks and producers demanding protection from foreign competition are likely to gain in importance. This will add a new dimension to attempts at international agricultural and trade policy coordination such as those in the GATT.

There has been a growing incidence of international disputes on food quality and food-quality standards. There was disagreement between the European Community and the United States over the use of hormones in beef production, and the Community threatened to ban the import of all U.S. beef for this reason.

In the real world, there are apparently impediments to a harmonization of food-quality standards that are difficult to overcome. This can be demonstrated by the experience in the European Community and in the United States. One of the central objectives of the internal market in the European Community, scheduled for completion by the end of 1992, is removal of all internal barriers to trade. It is now apparent that the Community will fail to adopt a uniform set of food-quality standards before the 1992 deadline. The United States has been a nation state for more than 200 years. Still numerous interstate barriers to trade remain. Many are due to differences in food-quality and related standards.

The growing importance of food-quality standards will increasingly affect international trade and policy. Issues related to food-quality standards are most likely to become the major issues in international agricultural trade and policy disputes. They will also be important items on the agenda of future attempts at international food, agricultural, and trade policy coordination. Therefore, we conclude that international discussion of the growing demand for food quality represent a very promising and important area of research in agricultural economics.

Acknowledgments

Research was supported in part by Agricultural Experiment Station Project No. 14-64, University of Minnesota, the North Central Experiment Station Regional Project NC-194, and by the DFG (German Research Association). The chapter is based in part on Harald von Witzke and Ian Sheldon, "The Growing Demand for Food Quality: Implications for International Trade and Policy." University of Minnesota, St. Paul, MN, 1990. The authors wish to thank Philip C. Abbott and C. Ford Runge for valuable suggestions and comments.

References

Ackerloff, G. A. 1970. "The market for 'lemons': Quality uncertainty and the market mechanism." *Quarterly Journal of Economics* 84:488-500.

Backhaus, H. 1989. "Okologische aspekte and sicherheitsfragen bei freilandanwendung von genetisch modifizierten organismen in der landwirtschaft." *Berichte uber Landwirtschaft*, Sonderheft 201, Hamburg, Germany: Parey.

Brown, J. 1988. "Implications of bovine somatotropin on the value of fluid milk quota in Ontario." *American Journal of Agricultural Economics* 70:1219-1224.

"Biotechnik Spezial." 1988. *Wirtschaftswoche* 37.

Falconi, C., and T. L. Roe. 1991. "A model of the demand and supply of health effects of food substances." In J. Caswell (ed). *Economics of Food Safety*. Amhearst, MA: University of Massachusetts, Department of Resource Economics.

Fallert, R. F., T. McGuckin, C. Betts, and G. Bruner. 1987. *bST and the dairy industry: A national, regional, and farm-level analysis*. Washington, D.C: Economics Research Service, USDA.

Hanf, C. H., and V. Wright. 1990. "The Quality of Fresh Food and the Agribusiness Structure." Armidale, NSW, Australia: University of New England.

Huisinga, R. 1989. "Technikfolgenbewertung: Bestandsaufnahme, kritik, perspektiven." *Leben und Erkenntnis von der Gesellschaft zur Forderung arbeitsorientierter Forschung und Bildung*. Germany: Frankfurt.

Juskevich, J. C., and C. G. Guyer. 1990. "Bovine growth hormone: Human food safety evaluation." *Science* 249:875-884.

Kalter, R. J. 1984. "Production cost: commercial potential and the economic implications of administering bovine growth hormone." *Proceedings of the Cornell Nutrition Conference for Feed Manufacturers*. Ithaca, NY: Cornell University.

Kelch, D. R. 1988. "EC livestock: Three EC actions threaten U.S. meat exports, and dairy quotas will continue." *Western Europe Agriculture and Trade Report: Situation and Outlook*. RS-88-1. Washington, D.C: Economic Research Service, USDA.

Kinsey, J. D. 1990. "Food quality and prices." *Agricultural and Food Policy for the 1990s*. Washington, D.C: Resources for the Future.

Kinsey, J. D., and J. P. Houck. 1991. "The growing demand for food quality: implications for international trade." In M. Shane and H. von Witzke (eds). *Public Goods in International Agricultural Trade*. St. Paul, MN: University of Minnesota, Department of Agriculture and Applied Economics. Staff Paper No. P91-12.

Klein, B., and B. Leffler. 1981. "The role of market forces in assuring contractual performance." *Journal of Political Economy* 89:615-641.

Kramer, C. S. 1990. "Food policy and public policy: What can economists contribute?" *Agricultural and Food Policy for the 1990s*. Washington, D.C: Resources for the Future.

Laird, S., and A. Yeats. 1990. "Trends in nontariff standards of developed countries 1966-1986." *Weltwirtschaftliches Archiv* 126:299-324.

Langbehn, C., and H. W. Wohlers. 1987. "Betriebswirtschaftliche Analyse des Einsatzes von Bovinen Somatotropin in der Milchkuhhaltung." Arbeitsberichte des Instituts fur landwirtliche Betriebslehre, Kiel.

Malthus, T. R. 1798. *An Essay on the Principle of Population as it Affects the Future Improvement of Society*. London, UK: Johnson.

Marion, B. W., and R. L. Wills. 1990. "A prospective assessment of the impacts of bovine somatotropin: A case study of Wisconsin." *American Journal of Agricultural Economics* 72:326-336.

Nelson, P. 1970. "Information and consumer behavior." *Journal of Political Economy* 78:311-329.

Organization for Economic Cooperation and Development. 1988. "Biotechnology and the changing role of government." Paris, France.

Ricardo, D. 1895. *Principles of Political Economy and Taxation*. New York, NY: McMillan.

Riethmuller, P., and T. L. Roe. 1986. "Government intervention in commodity markets: Japanese rice and wheat policy." *Journal of Policy Modeling* 8:327-348.

Runge, C. F. 1991. "International public goods, export subsidies and the harmonization of environmental regulations." In M. Shane and H. von Witzke (eds). *Public Goods in International Agricultural Trade*. St. Paul, MN: University of Minnesota, Department of Agriculture and Applied Economics. Staff Paper No. P91-12.

Runge, C. F., H. von Witzke, and S. J. Thompson. 1989. "International agricultural policy: A political economic coordination game." In H. von Witzke et al. (eds). *Policy Coordination in World Agriculture*. Kiel, Germany: Vauk.

Schmidt, G. H. 1989. "Economics of using bovine somatotropin in dairy cows and potential impact on the U.S. dairy industry." *Journal of Dairy Science* 72:734-45.

Senauer, B. 1979. "The effect of demographic shifts and changes in the income distribution on food-away-from-home expenditures." *American Journal of Agricultural Economics* 61:1046-1057.

Spinner, H. F. 1988. "Technikfolgenforschung im uberblick." *Der Hochschullehrer* 2.

Stiglitz, J. E. 1988. "Information and economic analysis." *Economic Journal* 95: Suppl. 21-41.

Witzke, H. von. 1986. "Endogenous supranational policy decisions: The common agricultural policy of the European community." *Public Choice* 48:157-174.

____. 1990. "Determinants of the U.S. wheat producer support price: Do presidential elections matter? *Public Choice* 64:155-165.

Witzke, H. von, and V. W. Ruttan. 1989. "Agricultural policies: A need for reform." *Economic Impact* 2:60-64.

Witzke, H. von, and I. M. Sheldon. 1990. "The growing demand for food quality: Implications for agricultural trade and policy." St. Paul, MN: University of Minnesota, Department of Agriculture and Applied Economics.

Zepeda, L. 1988. "The potential economic effects of bovine somatotropin on the California dairy industry." Ph.D. dissertation, University of California.

Zeddies, J., and R. Dolutschitz. 1988. "Potentielle einzelbetriebliche und sektorale auswirkungen des einsatzes von bovinem somatotropin (BST) in der milcherzeugung der bundesrepublik deutschland." *Berichte uber Landwirtschaft* 66:295-324.

PART FIVE

Policy Conclusions

14

BST: Issues, Facts, and Controversies

M. C. Hallberg

Historically, support for science-based technology has been viewed as the path to human betterment. Few in society questioned the wisdom of rapid adoption of the technology that science produced. This has been particularly true in agriculture. Johnson writing in the 1943-47 *Yearbook of Agriculture* captured the prevailing view toward technology in agriculture as follows:

> When men, as allies, use science to get the most from each acre, each worker, each machine, each animal, they make it possible for the earth or a part of it to support more people--and so to postpone indefinitely the day when the pressure of a population increasing faster than life-supporting resources would impel men to fight one another. . . . Scientific progress enables some people to live better, and more people to live. (Johnson 1947, p 920)

The record of productivity growth in agriculture during the present century is testimony to the fact that this sector has been extremely successful at adopting new technology. In 1900 there were 76 million residents in the United States. These residents were fed (and clothed) from a base of 1,151 million acres of crop and grazing land, and there was a sufficient surplus to export 29 percent (by value) of our agricultural production. Today there are more than three and one-quarter times that many residents in the United States being sustained from a base of only slightly more crop and grazing land (1,208 million acres), and we still produce an amount sufficient to export 25 percent (by value) of our agricultural production.

In dairy, where exports and imports have historically been minimal because of government policy, the evidence of technological advance has been even more striking. In 1900, one milk cow produced enough milk to satisfy the milk demands of 4.6 U.S. residents, whereas by 1990 one milk cow produced enough milk to satisfy the milk demands of 24.3 U.S. residents. For agriculture as a whole as well as for the dairy sector, most of the productivity gain brought about by new technology occurred after World War II. The research and extension arms of the Land Grant Universities of the nation have appropriately been credited with a major contribution for developing and implementing much of this technology.

Impacts of Biotechnology

As Lacy, Lacy, and Busch point out in the initial chapter to this volume, the technological revolution appears far from over. New technical breakthroughs are on the horizon promising additional productivity gains in agriculture. These authors see such gains coming from biotechnology developments in animals, plants, weed control, insect and disease resistance in plants, biofertilizers, and food processing. As with previous technologies, these hold great potential for both good and evil. Further, they can be expected to have impacts beyond the specific agricultural commodity affected--in the agricultural-input industries, in food processing, in the rural communities, and even in the public and private research sectors.

A recent Organization for Economic Cooperation and Development (1989) report identifies five conditions that must be satisfied for a new technological development or system to have a major impact on the economy as a whole:

1. A new range of products accompanied by an improvement in the technical characteristics of many products and processes, in terms of improved reliability, new properties, better quality, accuracy, speed or other performance characteristics.
2. A reduction in costs of many products and services.
3. Social and political acceptability.
4. Environmental acceptability.
5. Pervasive effects throughout the economic system.

For the authors of this report, nuclear technology fails on all five counts, whereas electric power and computer-based information technology satisfy all five criteria. Whether biotechnology, and in particular bovine somatotropin, the biotechnology product of concern in this book, satisfies all of these criteria is less clear.

There appears little doubt that the new biotechnology satisfies the first of these criteria. Bovine somatotropin is one evidence of this. The picture is somewhat more cloudy and difficult to unravel when we come to the second criterion, although it seems reasonably clear that BST, at least, will satisfy this one as well.

While the major focus of attention of those considering the impacts of new technology is often on the first two criteria, the last three are also important and must not be ignored. There may be long delays in social acceptance of radically new technologies such as biotechnology. Regulatory, educational, and managerial changes may be required. Changes in consumer attitudes and tastes may evolve and are often unpredictable. Primary environmental impacts of biotechnological developments do not appear great. Less obvious environmental consequences, however, may result from some of the secondary impacts of this technology. Since biotechnology operates through living organisms, most industrial sectors would be excluded from the direct influence of biotechnology. Similarly, BST will have its most immediate impact on the dairy sector. Indirect or secondary influences on other industries, however, cannot be excluded. Hence the technology may be quite pervasive as promises to be the case with BST. Some restructuring of parts of the national or world economy may accompany adoption of the new technology. All of these issues should be part of the process of assessing the new biotechnologies.

Costs of Technical Change

It is not always costless *to make two blades of grass grow where one grew before*. Further, conflicts can arise over technology adoption, as we have already witnessed in the case of BST. Indeed, as Johnson went on to warn:

> . . . history affords evidence that technological improvements, which bring profits to the producers who can adopt them, and which benefit mankind in general, also bring misery and distress to the individuals who cannot adjust themselves to the new conditions. Such individuals are likely to resist and may be strong enough to delay technological progress: Lancashire spinners and weavers, ruined by the invention of power spindles and looms, for

> instance, tried to destroy them before they gave up to become workers in the English textile mills. (Johnson 1947, p 920).

Johnson did not foresee other issues about which modern society would become concerned: pesticides that can be harmful to the environment and are toxic to man and animals; fertilizers that lead to pollution of underground and surface water; machines that not only displace farm workers but also disrupt social structures; and a combination of mechanical, chemical, biological, and managerial developments or aids that lead to a reduction in farm numbers, threaten to diminish the quality of rural life, increase the capital necessary for new entrants into farming, and impose unforeseen costs on future generations.

Society does care about these issues and is, apparently, ready to do something about them. In 1900, consumers allocated as much as 40 percent of their total personal-consumption expenditures for food. Today only about 17 percent of the total personal-consumption expenditure is allocated for food. Thus consumers are not only willing to sacrifice some efficiency in food production for other issues they deem important, many feel they can *afford* to do so. As Robert Persig in *Zen and the Art of Motorcycle Maintenance* put it:

> . . . What's wrong with technology is that it's not connected in any real way with matters of the spirit and of the heart. And so it does blind, ugly things quite by accident and gets hated for that. People haven't paid much attention to this before because the big concern has been with food, clothing and shelter for everyone and technology has provided these. But now where these are assured, the ugliness is being noticed more and more and people are asking if we must always suffer spiritually and esthetically in order to satisfy material needs. (Pirsig 1974, p 149).

We have, thus, moved into an era where ethical issues need to be faced head on. BST, the major topic of this volume, clearly does involve ethical issues, as Thompson points out in Chapter 2. These ethical issues arise from four sources: (1) impacts on dairy farmers and rural communities, (2) perceived consequences on dairy cows, (3) environmental impacts of projected intensification of dairy production, and (4) uncertainty concerning the safety of milk and its products produced from cows treated with BST.

Some might argue that scientists have contributed to, rather than diffused, the contentious issues surrounding BST. Scientists are specialists, and are in general very knowledgeable in their area of specialty. As specialists, scientists tend to operate in a very small world, and when asked to travel outside their small world they sometimes give either half-truths or are hesitant to react at all. When a social scientist, for example, attempts to review the animal health consequences of BST, less than the complete story can be expected. The social scientist is not likely to be familiar with all of the animal health literature nor even with the technical terms used in the animal health professions. Similarly when an animal geneticist is asked to comment on the economic and/or social consequences of BST, the questioner should generally be wary. It is difficult enough for the geneticist to keep up in her or his own field, let alone keep up with past and ongoing research in the social sciences. This is the nature of science. There is probably no better way.

Unfortunately, all this is unsettling to the public, and provides an opportunity for controversy if not misrepresentations. When scientists from different specialties relate conflicting ideas concerning issues outside their area of specialty, the public becomes confused or even irritated, and loses trust. The media and/or special interest groups then enter the fray. The latter often contribute to the confusion, not because they aim to confuse or mislead, but because they lack the knowledge with which to assimilate and interpret the research that is available. Others, no doubt, play on this confusion to promote a particular point of view. There is, then, no substitute for clear and concise statements from all relevant perspectives and disciplines as to what is known and what is unknown about an issue such as BST. This is the spirit in which this book project was undertaken.

Assessing the Consequences

Referring back to the ethical issues identified by Thompson, one might well ask, what is new and/or significant in the case of BST? Cast in the historical context of productivity growth in dairy production, is BST likely to be more significant than any of a number of other technologies that have been adopted since, say, World War II? Will BST bring about a reduction of dairy cows at the rate witnessed between the census years of 1969 and 1987--a loss of more than 1 million dairy cows for a compound annual loss rate of 0.33 percent? Will BST bring about a loss of farm operations with milk cows at the rate witnessed between 1969 and 1987--a loss of more than 360,000 operations for a compound annual loss rate of 3.63 percent? (U.S. Department of Commerce 1989). Will there be significant animal health problems associated with the use of BST? Will the environmental consequences be more severe than those we have sustained to date? Indeed, will there be a significant intensification of dairy operations as a direct result of BST use? Is there cause for consumers to be more concerned about the wholesomeness and safety of milk from BST-injected cows than of fortified, homogenized, and pasteurized milk?

To be sure, there is yet much to be learned about BST use. We do not know how difficult it will be to overcome the managerial complexities likely to be attendant to BST use (Patton and Heald, Chapter 5). We can not know how rapidly farmers will adopt BST until it becomes commercially available (Yonkers, Chapter 7). We will also not know how rapidly and to what extent milk prices will adjust downward (Fallert and Hallberg, Chapter 10) until we know how many dairy farmers adopt this technology. Finally, we will not know how many consumers will reduce their consumption of milk and dairy products until milk produced from cows injected with BST is generally available in the market place (Smith and Warland, Chapter 11). Indeed, we may *never* know if consumers who reduce their consumption of milk and dairy products do so because they fear the product is unsafe for human consumption, or because they wish to register their concern about the unwanted, perceived consequences of BST.

Nevertheless, there is much that we do know about BST and its likely consequences as the various authors of this book have documented. Viewed from a historical perspective and based on the research evidence presented here, it would appear that all of the questions posed above should be answered in the negative. In general, this book provides much of the detail needed to answer many fundamental questions concerning BST. This, however, is *not* to say that no unwanted consequences will accompany BST use nor that no uncertainties remain. BST use will certainly intensify, not ease, past trends in agriculture about which there has been continual unrest: a lower ratio of prices received by farmers to prices paid by farmers, adjustment of labor out of agriculture, and fewer and larger commercial farms (Tauer, Chapter 9). Further, BST will have an impact on related industries and markets (Greaser, Chapter 8).

Policy Approaches

A key policy question is what to do about the unwanted consequences of BST and/or about the uncertainties that remain. Some argue that the unwanted consequences and uncertainties associated with BST are so severe that the product should be banned. Indeed legislation has been introduced in Minnesota and Wisconsin to extend the existing ban on the commercial sale of BST milk in those states. Some organizations threaten to mount a nationwide boycott of BST if the Food and Drug Administration approves the product. The European Community has declared a moratorium on BST use through 1991, and four Scandinavian countries have banned BST. (The European Community has recently found Monsanto's version of BST approvable for commercial use, but the current moritorium on BST use in the European Community remains in effect.) All but one province in Canada have terminated BST trials (the exception is Quebec) pending governmental approval of the product, but there has been no legislative action seeking a moratorium on BST use in Canada. At the same time several countries have approved the commercial sale of BST: Mexico, Brazil, the USSR, Bulgaria, Czechoslovakia, South Africa, Namibia, Malasia, India, Pakistan, and Zimbabwe. A host of countries have authorized milk from BST-treated cows to enter the food chain and several have BST research efforts underway.

Roush (1991) offers an approach for dealing with BST-like technologies that would, hopefully, reduce the controversy that BST has stirred up and prevent the unwanted consequences that BST threatens. This approach would involve laypeople at the beginning of the publicly funded research and development process so that *inappropriate* technologies and products can be aborted before they acquire too much momentum. Another approach, again suggested in the article by Roush (1991), would be to add a *fourth hurdle* to the FDA-approval process: considering a new product's social and economic effects (as does the European Community), as well as its safety, efficacy, and quality as does the Food and Drug Administration in the United States.

None of these options is particularly attractive as a mechanism for dealing with BST or for what they imply about the potential reaction to future technologies. Even in the absence of BST, a good bet is that the annual compound growth rate in milk output per cow between 1970 and 1990 of more than 2 percent will be sustained into the foreseeable future as a result of (1) continued genetic improvements in the dairy cow, (2) improved dairy feed additives, and (3) managerial improvements leading to increased efficiency in dairy production. Since the latter will lead to the same *unwanted* consequences as will BST, does the banning of BST mean that we should also ban all future new or genetic improvements in dairy animals, all future improved feed additives, and all future improvments in management? Certainly, prohibition of future technological change that threatens some social and economic disruption is inconsistent with our general objectives of economic growth and progress. At the same time, however, it might well be said that the scientific community bears some responsibility for the social and economic disruptions associated with technical change.

One of the implicit performance criteria by which the agricultural sector has in the past been judged is its capacity to adopt new methods and practices so as to maintain its competitive edge and to provide consumers the benefits of cost-reducing technology. The U.S. agricultural sector must receive high marks for its efforts in this area, in spite of sustained government efforts to curtail production. If one subscribes to this performance criterion, then one must conclude that a most appropriate technology policy response for agriculture is to encourage adjustments so that resources (both human and physical resources) are employed in their best use. Elsewhere I have argued that assets are not as *fixed* in agriculture as some would have us believe (Hallberg 1989). Nevertheless, there are instances in which adjustments are difficult because assets are somewhat fixed in the short run, farm people do not have the skills needed for alternative employment, or alternative employment within or outside of agriculture is not readily available. If this is the case, a rational approach would be to develop policies aimed specifically at easing the adjustment process. This may take the form of job retraining, development efforts designed to create nonfarm jobs, farm-asset buyout programs, developing new viabilities in agriculture, etc. All this argues for continuing to strive for an economy that provides a wide range of alternative opportunities for those displaced by the type of technology under discussion here. This would contribute not only to general economic growth and progress, it would lessen the imperative to provide specific protection from a given technology.

This policy response can equally be applied to the case of BST-like technologies and is apparently what Bromley (1991) had in mind when he offered his *policy for transitions*. Bromley suggests that:

> . . . if the issue is perceived as one of easing the private and social costs of the transitions from one technological path to another, then the idea of a technology policy seems more consistent with traditional economic policy. Certainly all economically disruptive technological change cannot be prohibited. But, it is equally clear that the lifestyle and economic security of thousands of people and their families are too important to be dismissed as irrelevant in the face of some new technical opportunity--particularly when that new opportunity promises to boost the income of only a very few able to promote its adoption. . . . In agriculture, family farm advocates who feel threatened by biotechnology would probably support a public streategy that would anticipate and recognize threats to traditional modes of production, and they, in response, would undertake research and public education programs to assist with problems of economic transition.

Acknowledgments

I am grateful to Professor Donald Crider for comments and suggestions on an earlier draft of this chapter.

References

Bromley, D. W. 1991. "Technology, technical change, and public policy: The need for collective decisions." *Choices*. Second quarter.

Hallberg, M. C. 1989. *The U.S. Agricultural and Food System: A Postwar Historical Perspective*. Publication No. 55. University Park, PA: The Northeast Regional Center for Rural Development, The Pennsylvania State University.

Johnson, S. E. 1947. "Farm Science and Citizens." Pp 920-932 in U.S. Department of Agriculture. *Science in Farming: Yearbook of Agriculture, 1943-1947*. Washington D.C: U.S. Government Printing Office.

Organization for Economic Cooperation and Development. 1989. "BioTechnology: Economic and Wider Impacts." Paris, France.

Pirsig, R. 1974. *Zen and the Art of Motorcycle Maintenance*. New York, NY: Bantam Books.

Roush, W. 1991. "Who decides about biotech? The clash over bovine growth hormone." *Technology Review* 94:5:28-37.

U.S. Department of Commerce. 1989. *1987 Agricultural Census*. Washington, D.C: Bureau of the Census, U.S. Government Printing Office.

About the Contributors

L. J. (Bees) Butler is professor of agricultural economics extension at the University of California, Davis. Dr. Butler has done extensive research on the Plant Variety Protection Act and the U.S. seed industry. He has been a member of the council for Agricultural Science and Technology task force on Germ Plasm Preservation and Utilization in the United States. More recently he has been involved in various economic assessments of BST use both in the United States and in foreign countries.

Lawrence Busch is professor of sociology at Michigan State University. He is coauthor or coeditor of six books on agricultural research policy the latest of which is entitled *Plants, Power, and Profit: Social, Economic, and Ethical Consequences of the New Biotechnologies* (Basil Blackwell, 1991). Currently, he serves on the scientific advisory board of CIRAD, the French agricultural scientific cooperation agency. Dr. Busch's interests include agricultural science and technology policy, higher education in agriculture, and public participation in the policy process.

Frederick H. Buttel is professor of rural sociology and science and technology studies at Cornell University. He is particularly interested in the socioeconomic aspects of biotechnology and agriculture in university-industry relationships, in agribusiness reorganization, and in the impacts of biotechnology on Third World agriculture. He served as President of The Rural Sociological Society during 1990-91. Dr. Buttel's numerous consultancies include the Ford Foundation, The Congressional Office of Technology Assessment, the Institute for Alternative Agriculture, the National Rural Center, and the National Council of Churches.

Richard F. Fallert is senior economist in the Livestock, Dairy and Poultry Branch, Economic Research Service, USDA. For ten years he served as Leader of the Dairy Research Section of that Branch. Over his career with Economic Research Service, Dr. Fallert has been involved in a variety of economic assessments relating to policy options for the U.S. dairy industry, and in studies concerning the overall structure and growth of the U.S. dairy industry. More recently his research interests have included economic assessments of BST adoption for which he has received commendations from ERS.

George L. Greaser is an extension economist at The Pennsylvania State University with a joint appointment in the Department of Agricultural Economics and Rural Sociology and Dairy and Animal Science. Prior to joining the faculty at The Pennsylvania State University he was an associate professor of agricultural economics at Colorado State University. Dr. Greaser has participated in several regional farm management and dairy situation conferences. Throughout his career he has played a lead role in developing and updating cost of production information for various agricultural enterprises. Currently he is actively involved in the development of computerized dairy management tools using various decision-making techniques.

Milton C. Hallberg is professor of agricultural economics at The Pennsylvania State University. Much of his past research has been concerned with national policy for the U.S. dairy industry. His current interests include, in addition to dairy policy, technological change and adjustments in agriculture, linkages between the farm and non-farm economy, and international agricultural trade and policy. He has participated on various national committees and Task Forces including the American Agricultural Economic Association Task Force on Dairy Marketing Orders. He has served as Economist with Economic Research Service, USDA, and as Visiting Professor to Universities in the United States, Australia, West Germany, and Northern Ireland.

Claus-Hennig Hanf is professor of agricultural economics and chair of the Institute of Farm Management and Production Economics at The Christian-Albrechts-University of Kiel in Germany. He holds the Ph.D degree from the University of Sluttgart-Hohenheim where he taught before coming to Kiel. He has also served as Dean of the Faculty of Agricultural Sciences at Kiel. He has held several visiting scholar posts at universities in Austria, Australia, France, Sweden, and the United States. He was President of the European Association of Agricultural Economists in 1985-87. Dr. Hanf's research interests have ranged over a wide group of subjects, in production economics, firm decision analysis, and sector modeling. He has published widely in these areas and more recently has been actively engaged in marketing and policy analyses of the German and EC dairy industries.

C. William Heald is professor and extension coordinator for dairy and animal science at The Pennsylvania State University. His special areas of interest include somatic cell count methods, mastitis control, and production record management. He is a contributor to a textbook used in courses on lactation physiology, co-author of a chapter in a four-volume treatise on lactation physiology, and author of several fact sheets in the National Cooperative Dairy Herd Improvement Program Handbook. He has served on several committees of the Dairy Science Association and currently serves on the Editorial Board of that Association. He was one of several Extension Dairy Specialists organized nationally to develop core parameters for use by dairy farm managers in the Dairy Herd Improvement Association program.

Lawrence J. Hutchinson is professor of veterinary science at The Pennsylvania State University. His areas of research interest include paratuberculosis infection in cattle, tuberculosis infection in swine, bovine mastitis, and bovine reproductive efficiency. After fourteen years in private veterinary practice, he became extension veterinarian at The Pennsylvania State University. His extension emphases are on health and productivity of dairy cattle and continuing education programs for veterinarians.

Manfred Kroger is professor of food science at The Pennsylvania State University. He maintains an active interest in food chemistry, food laws and regulations, food toxicology, environmental toxicology, and dairy technology. He is a Regional Communicator for the Public Information Program of The Institute of Food Technologists, an advisor to the American Council on Science and Health, and a consultant to the International Food Information Council. He is currently associate editor of the *Journal of Food Science*.

Laura R. Lacy is a research scientist in the Department of Plant Pathology at The Pennsylvania State University. Dr. Lacy's current research interests are in the areas of plant response to abiotic and biotic stresses, and the application of biotechnology to crop improvement. She recently coauthored *Plants, Power, and Profits: Social, Economic, and Ethical Consequences of the New Biotechnologies* (Basil Blackwell 1991).

William B. Lacy is Assistant Dean for Research in the College of Agriculture and professor of rural sociology at The Pennsylvania State University. He has had a long-standing interest in the agricultural research process and in research administration in agriculture and has co-authored or co-edited *Science, Agriculture, and the Politics of*

Research (Westview Press 1983), *Food Security in the United States* (Westview Press 1984), *The Agricultural Scientific Enterprise* (Westview Press 1986), and *Plants, Power, and Profit: Social, Economic, and Ethical Consequences of the New Biotechnologies* (Basil Blackwell 1991).

Dale A. Moore, MS, DVM, MPVM, is field investigator and coordinator of the Dairy Production Medicine Program at the Department of Veterinary Science, The Pennsylvania State University. After two years in private practice in California, three years as a resident in Food Animal Medicine and Herd Health at University of California Veterinary Medical Teaching Hospital, and two years as an epidemiologist for the Federal Centers for Disease Control, she joined The Pennsylvania State University in 1990. Her areas of interest include: bovine disease and reproduction and dairy production medicine.

Lawrence D. Muller is professor of dairy science at The Pennsylvania State University. Dr. Muller's research interests are in the areas of dairy nutrition and management with emphasis in forage evaluation. He has been an active participant in the American Dairy Science Association, serving on several different committees including the editorial board of the *Journal of Dairy Science* and the Board of Directors. He was also appointed to a committee of seven to rewrite the National Research Council's *Nutrient Requirements of Dairy Cattle*.

Robert A. Patton is a technical specialist on Monsanto's BST field staff and has been involved with clinical farm trials in several states. Educating farmers, veterinarians, dairy consultants, and sales representatives on the use of BST as a management tool is also part of his job. He has spent many weeks in Mexico, where BST is on the market, providing technical support and observing the results of BST usage in a variety of farm management systems.

Dr. Patton has a masters degree from The Pennsylvania State University and a doctorate from Michigan State University, both in ruminant nutrition. He has been a dairy farmer, a state extension specialist in dairy nutrition, a county extension agent, a college instructor of dairy herd management, and a private nutrition consultant. His research interests are in feed intake and the effects of management practices on milk production.

Blair J. Smith is professor of agricultural economics at The Pennsylvania State University. Throughout his career at The University of Florida, The University of Georgia, and The Pennsylvania State University, Professor Smith has devoted his major efforts to research on the economics of dairy production, marketing, and policy. More recently he has concentrated his efforts on component pricing, market concentration, and consumer attitudes and behavior toward drinking milk.

Loren W. Tauer is associate professor of agricultural economics at Cornell University. His research has involved a variety of issues in the dairy production area, including the structural and economic consequences of dairy technologies. He is currently editor of the *Northeastern Journal of Agricultural and Resource Economics.*

Paul B. Thompson is associate professor of philosophy and of agricultural economics at Texas A&M University. His teaching and research responsibilities include social and political philosophy, the philosophy of the social sciences, and ethical issues in agricultural policy. He was an International Affairs Fellow with the Council on Foreign Relations during 1986-87 and held a visiting scholar post at the U.S. Agency for International Development where he conducted research on the philosophical conflict between international development in agriculture and the commercial interests of American farmers.

Rex H. Warland is professor of rural sociology at The Pennsylvania State University where he has been actively involved throughout most of his career in studying consumer food choice behavior, consumer reaction of imitation foods, nutrition as a factor in food choice behavior, and the consumer movement. Professor Warland is noted also for his work in evaluation methodology and measurement. He has served as associate editor of the Journal of Consumer Affairs and is a member of the Governing Council of the Rural Sociological Society.

Robert Yonkers is assistant professor of agricultural economics at The Pennsylvania State University. He has statewide responsibilities to develop and implement extension and research programs in dairy marketing and policy. Before joining the The Pennsylvania State University faculty, he was staff economist in the Agricultural and Food Policy Center at Texas A&M University where he obtained his Ph.D. degree.

Harald von Witzke is associate professor in the Department of Agricultural and Applied Economics, University of Minnesota and serves as Director of the Center for International Food and Agricultural Policy. He holds a Ph.D. in agricultural economics from the University of Goettingen, where he taught before coming to Minnesota. He also served with the Ministry of Agriculture in

Germany and taught at the University of Bonn. Dr. von Witzke specializes in agricultural development, policy, and trade. He has a special interest in the European Community, Eastern Europe and U.S.-European agricultural and trade policy relations.

Glossary

Adipose tissue: Fat-containing tissue in the living organism.

Amino acid: Any of a group of 20 unique chemicals that link together in chain-like combinations to form specific protein molecules in the cell.

Biotechnology: Any technique that uses living organisms or substances to make or modify a product, to improve plants or animals, or to develop micro-organisms for specific uses. These techniques include the use of recombinant DNA technology, cell fusion, and other bioprocesses.

Body condition score (BCS): A measure (on a scale of 1 to 5) of stored fat reserves in the cow.

Bovine somatotropin (BST): A polypeptide (protein) hormone, consisting of a chain of 191 amino acids, produced naturally by the anterior pituitary of the cow. Its function is to apportion nutrients toward milk production. It can also be produced commercially by recombinant-DNA technology and injected into cows. The BST product manufactured by Upjohn is identical to pituitary BST, while that manufactured by American Cyanamid contains three additional amino acids and that by Eli Lilly nine additional amino acids. Monsanto's BST has one amino acid substitute.

Cell: The smallest component of life, and the fundamental building block of all living organisms. Some organisms (e.g., *Escherichia coli*) are unicellular (meaning that the cell itself is a complete organism). The higher plants and animals, however, are made up of hugh numbers of mutually dependent cells that perform specialized functions e.g., growth, infection control, or reproduction). Each cell controls and performs a myriad of chemical reactions, extracting and applying energy from molecules to the support of life processes for itself and for the organism of which it is a part.

Cell-culture techniques: The propagation and maintenance, under laboratory conditions, of unicellular organisms or of individual cells derived from multicellular organisms for study or for manufacturing purposes.

Chromosome: Bodies within the cell where hereditary information (genes) are located. Chromosomes are composed primarily of DNA material.

Class I milk: Grade A milk utilized by processors, regulated under milk marketing orders, to produce fresh fluid milk products.

Class II milk: Grade A milk utilized by processors, regulated under milk marketing orders, to manufacture soft dairy products such as cottage cheese, ice cream, and yogurt.

Class III milk: Grade A milk utilized by processors, regulated under milk marketing orders, to manufacture hard dairy products such as butter, cheddar cheese, and nonfat dry milk.

Classified pricing of milk: The system of pricing milk according to the use made of that milk by processors regulated under milk marketing orders. In most marketing orders, three use classes of milk are established: Class I, Class II, and Class III milk. For analytical purposes, Class II and Class III milk are generally treated as the same and referred to collectively as Class II milk since both are used to produce manufactured products.

Deoxyribonucleic acid (DNA): Also called deoxyribose nucleic acid. A nucleic acid found in all living cells that is the carrier of all genetic information for the living organism. DNA consists of two long chains of alternating phosphate and sugar (deoxyribose) units twisted into a double helix and joined by hydrogen bonds.

Enzyme: A specialized protein molecule that coordinates, or acts as a catalyst for, chemical activities within the cell.

Escherichia coli: A single-celled bacterium (and thus a complete organism) that is a common inhabitant of the human gut. All *E. coli* cells carry the same DNA, produce the same proteins, and carry out identical biochemical reactions. They are easily cultured and, thus, frequently used in biotechnology applications.

Estrous cycle: The period of time needed for the reproductive cycle in the female, nonhuman mammal that includes egg maturation and ovulation, and preparation of the uterus to receive fertilized eggs.

Estrus: The period during which a female, nonhuman mammal is receptive to sexual activity.

Fluid milk: Producer milk used for producing fresh fluid milk products.

Gene: The fundamental unit of heredity. Each gene contains DNA which encodes information that defines the structure of a specific cell or causes the cell to perform a specific function.

Genetic engineering: Technology (including recombinant DNA technology) used by scientists to isolate genes from an organism, manipulate them in the laboratory and insert them into another organism.

Genome: A complete set of chromosomes representing the total genetic structure of a cell.

Genotype: The total genetic constitution contained in the chromosomes of an organism as distinguished from that organism's physical appearance or phenotype.

Germ cells: Sperm and ova, the cells involved in fertilization and reproduction. Germ cells have one-half the full complement of chromosomes.

Grade A milk: Producer milk that meets federal sanitary standards for processing into fresh fluid milk products.

Grade B milk: Producer milk that fails to meet federal sanitary standards for production of fresh fluid milk products. All such milk must be used to produce manufactured dairy products.

Homeorrhesis: The genetically controlled tendency in the living organism to maintain a particular biological process, such as growth, along a specific path despite factors tending to divert it away from this path.

Hormone: A chemical substance of fatty-acid origin produced by one of the endocrine glands that has a specific function or produces a specific effect relating to growth, metabolism, or reproduction.

IGF-I: A 70-amino acid polypeptide insulin-like growth factor. This is the major such factor serving to mediate the effects of a growth hormone.

IGF-II: A 67-amino acid polypeptide insulin-like growth factor. This factor plays a less dominant role in mediating the effects of a growth hormone than does IGF-I.

Insulin: A hormone secreted by the pancreas into the blood that regulates sugar metabolism.

Interleukin 2 (IL-2): A mediator acting upon the lymphocytes enhancing their capacity to respond to antigens.

***In vivo*:** Within the living organism.

***In vitro*:** Outside the living organism and in an artificial environment.

Lipolysis: Hydrolysis of fat into free fatty acids and glycerol.

Luteal activity: Hormone producing activity by the corpus luteum on the ovary.

Manufacturing milk: Producer milk used to produce manufactured dairy products.

Mastitis: Inflammation of the udder leading to reduced milk output and to a reduction in the protein content of milk produced.

Multiparous: Having given birth more than once.

Neutrophil: A white blood cell easily stained by neutral dyes.

Peripartum: Around or about the time of giving birth.

Parturition: The act of giving birth.

Phagocytosis: The process of cell digestion of bacteria and other foreign bodies in the bloodstream and tissues.

Phenotype: The environmentally and genetically determined physical appearance of a living organism.

Phytoproduction: Production by plants.

Phytohemagglutin (PHA): An antibody that stimulates lymphocyte proliferation.

Polypeptide: Any natural or synthetic compound (protein) containing 10 or more amino acids.

Postpartum: After having given birth.

Postparturient: Having given birth.

Primiparous: Having given birth once.

Progesterone: A hormone secreted by the corpus luteum (or prepared synthetically) serving to prepare the uterus for the reception and development of the fertilized ovum.

Prostaglandins: A hormone which causes the normal estrus cycle to be shortened.

Recombinant-DNA technology: A series of genetic-engineering techniques for manipulating DNA.

Somatic cell: Any cell of the body other than a germ cell. Somatic cells have a full complement of chromosomes. An unusually high number of somatic cells in milk are evidence of mastitis.

Somatic cell count: A count (estimate) of the number of somatic cells in milk. It is an indirect measure of the incidence of mastitis in cattle.

Tissue-culture techniques: The propagation and maintenance, under laboratory conditions, of tissue derived from multicellular organisms for study or manufacturing purposes. As distinguished from cell-culture techniques, here the cell grows as tissue material rather than as an independent cell.

Transgenic plant or animal: A plant or an animal whose hereditary DNA has been augmented by DNA from a source other than parental germplasm (i.e., from recombinant-DNA technology). Scientists have developed transgenic dairy animals that secrete valuable human proteins in their milk, and transgenic plants that possess *Bacillus thuringiensis* genes, providing insect resistance.

Virus: An infectious agent of a host cell consisting of nucleic acid and genetic material encapsulated in a protein shell. It survives outside the cell but cannot reproduce until it invades a living host cell where it can replicate with genetic continuity and the possibility of mutations.

Author Index

Subject Index